MARS II

by

T. D. Hudson

Mars II

Introduction

Mars II, a planet designed for the purpose of final conflict resolution, created by the Warrior Program which is a regulatory agency administered by United Planets Coalition. It consisted of more than 1000 Delegates from 305 planets, colonies, subsystems and species within in the Coalition. Their purpose was to create rules to govern types of weapons, areas are to be utilized, and who is assigned battle space at any given time. In essence the game of War has evolved into ... A Game of War.

Though the list of rules is endless, the basics for humanoids are:

1. Teams of four plus one neutral person (Tech) to act a liaison between referee's and other members. Only the neutral member may speak to the referee, exception being if the neutral member is killed or otherwise incapacitated. However, the neutral member is not a combat operative and should never be placed in harm's way.

2. Short range weapons only.

3. All battles to be fought within a given zone no larger than ten square kilometers.

4. No Weapons of Mass Destruction.

5. All opponents must be dead or incapacitated beyond their ability to fight further and verified so for the War to be ruled concluded.

6. All combatants must be approved by the Warrior Program before they can compete.

7. All combatants must complete the required training program to be considered for competition.

8. All combatants must volunteer and cannot be coerced into

battle.

9.   No combatant will be paid for services. Beyond the contracted amount agreed to at the completion of battle. Final payment cannot exceed 100.000 credits plus medical costs.

10.   No one may be excluded from consideration provided they meet the accepted minimum human age of consent set at 25 years or equivalent are of sound mind and body and are affiliated with the Planet they are competing for either by birth or allied affiliation. All combatants from each system must be approved by the UPC Directorate.

11.   Combatants can be recruited from any location but must recruit from their home world first and be given first consideration paid proxies are permitted provided such payment does not exceed the rate shown in rule 9.

12.   Teams may be constructed from allied planets if the decision to do so it made prior to scheduling on Mars II.

Earths recruits for the Warrior Program are taken from the military or on occasion maximum security prisons. Though unusual, it is not unheard of that the more violent offenders are the ones sought and recruited for the program. We begin on the Earth prison asteroid MM126, 600,000 kilometers from Martian Colony 21, affectionately known as

Outlaw 9...

And so it begins...

Jarob Ka was waiting for the ref's decision before he could administer the coup-de-gras on the opponent of his final match. His tri-saber lifted high and his arm outstretched he looked in defiance at the creature at his feet:

"The winner, Jarob Ka, I.D. number 100967-29A." came announcement.

"Yeah...yeah I know that, just get on with it." he mumbled.

"The disposition honor is yours." the referee continued.

The creature that now lay before Jarob wasn't human, though from all outward appearances it may have looked that way. He glared one last moment into those defiant red eyes his heart cold, no need for emotions. This wasn't a man, it was a thing, a number, a miserable piece of shit that had hawked him since he first arrived. Now, the only thing he could do was kill it, end its misery, and do it a favor. Somehow that wasn't enough, slow torture perhaps, or maybe do to it what it had tried several times to do to him. Or he could allow it to live.

"Fuck that." he yelled thrusting the tri-saber deep into the things cold, evil heart, twisting the saber roughly several times. The creature let out a final echoing cry as its eyes rolled white, then... silence as it lay bleeding into the sand-covered arena floor.

Jarob had been in this shit-hole five long years. Had put up with this piece of shit trying to punk him, steal his rations, and kill him at every turn. But now Jarob Ka, number 100967-29A was about to become warrior Jarob Ka, with no fucking number and that felt good. After training years, here in this asshole at the end of space, he had finally made warrior class. The first few years he was on the Outlaw, he did nothing but smoke cigarettes and anything else available. Sit on his ass in his cell, and avoid being jumped by the puke that now lay dying in a pool of his own blood in the arena, or others like it.

Sentenced to life here on the Outlaw was better than the two years he'd spent in other dumps back home. Pressure dome life wasn't so bad, some guys can't take it, go off their nut in a week or two. Those are the one's who usually off themselves, or sell their ass to the highest bidder. Either way they don't last long. His fuck you attitude is what got him here in the first place, so he figured he might as well make good use of it. Some squab of a kid tried sticking him up for a couple mags of Pyresium, dumb ass kid---Dead kid.

They called it murder he didn't see it that way. Still his attorney convinced him to cop to life, instead of death by whatever. Turns out the kid was some big wigs son, so rather than call it what it was, they called it murder and pinned the max on Jarob. Either way he was sure that somewhere down the line someone else got the credits for the Pyresium. He suspected it was probably his attorney and the big wig.

Like all prisons, the food was shit, the beds were hard, and the hum of the force field kept him awake the first few nights, but he adapted he was good at that. Jarob had heard the warrior program was going to recruit from the Outlaw from a supply clerk who heard about it from a shuttle pilot. So he sent in his paperwork and waited even though he knew the prison grapevine was usually shit. Someone got it right this time and he was recruited a couple of months later.

Jarob now stood in the arena next to the dead pile of sub-human garbage just one of many he had to fight on his rise to the top. Though it usually ended with someone dying, sometimes it didn't and most of the time he didn't care either way. He knew most of the men he fought some he liked, respected, and felt bad about what he had to do; this time it was different. This time it was gratifying, and though no spectators were allowed in the arena, he held the tri-saber aloft in a gesture of triumph.

"Yeah...ok, you won. Co'mon it's time for chow." a guard said flatly.

Jarob kicked sand in the dead things face as he left the arena.

Chow didn't taste as lousy as it usually did that evening even though it was the same rat stew and stale hard moldy bread. As he walked back to his cell the entire block was strangely quiet with none of the usual cat calls, boos, hisses, or "Holy shit, HE won!"

"Now what?" Jarob asked the guard.

"You'll get your orders in a few days, just sit tight." The guard replied, motioning him into his cell.

Jarob couldn't sleep; still keyed up.

"Hey Ka!" a voice yelled from a few cells down.

"Yeah...who is it." Jarob yelled back.

"S'me...Trimmic...D'you take that asshole down hard?"

"Damn Straight!" Jarob replied.

"Good...I hope you pig stuck him once for me."

"I made good on a lot of shit today."

"Look's like you'll be leavin' the Nine for greater glories on Mars II just watch your ass...son...g'night."

"Yeah...yeah I'll do that. Thanks...g'night."

Jarob wasn't sure how to react to Trimmic most of the time he got the impression Trimmic hated him. That is until recently when this match was announced, and Trimmic suddenly found his tongue telling Jarob in no uncertain terms to get that fucker.

Jarob waited six long goddamn days before the guard arrived with his orders. Excited as a nine year old on Christmas morning, he tore into the envelope. Enclosed were a waiver card, exempting the Warrior program from any problems that may occur while Jarob was in the program, an organ donor card, a pay voucher for 1000 credits, and the following form letter:

Dear Mr. Ka:

Congratulations on your recent admission to the Warrior program. Mr. Arthur Malcomb, one of our representatives, will be in contact with you shortly. He will have all the necessary paperwork in hand, to transfer you to our branch facility training center on Triton. For the next ten weeks thereafter, you will participate in a rigorous training program, and be placed on a rotation list for service on Mars II. Your commission will be that of Major while in our service. Your actions have proven you to be the best warrior on MM127. Thank you for your participation, we are sure you will make a fine addition to our team and the United Coalition of Planets bids you Welcome.

Anthony Cloudrock Sr.

Regional Director

Warrior Program

Jarob had just finished the letter when a man in a blue pinstriped suit appeared at the doorway to his cell.

"Mister Ka?" he said looking at the number over Jarobs door.

"Yeah."

"I have your exit papers here....."

"Watch..." Jarob said trying to warn the man about walking into the security grid in his cell door.

The puzzled look on the man's face as he flew backwards, slamming hard against the hallway wall made Jarob's eyes tear as he tried to contain his laughter.

"...out." Jarob continued doing his best to control his need to laugh himself silly.

After a few minutes in the infirmary, both he and Jarob were taken back to Jarobs cell to get a few personal items. Not much he really wanted to take with him anyway, he must have taken 10 showers the past six days in anticipation, and shaved so often he didn't even have stubble.

"You a...you been doin' this long?" Jarob asked with a broad smile.

"Fifteen years come May." the man replied extending his hand. "Oh a...Malcomb, Arthur Malcomb not a warrior, just an administrator."

"Well Malcomb Arthur Malcomb, fifteen years should have taught you, to tell the guards to deactivate the security screens, before you try to enter a cell. But, like you said you're just an administrator." Jarob replied taking his hand.

"S'first time, first time that's happened. I guess I was in too big of a hurry, I got three more pick ups to make this week within a sixty parsecs. Pressure, you know? It's just Arthur Malcomb by the way."

"Yeah...pressure. Ok, Just Arthur Malcomb, what next, we go to Triton...right?

"Yeah...a...sure, a...you got a mental or dyslexic disorder we should know about."

"Don't think so, why?"

"My name is just, plain, old, Arthur Malcomb, you got that."

"Shit...you're playing with my head now, first it was Malcomb Arthur Malcomb, then Just Arthur Malcomb, now its Just Plain Old Arthur Malcomb. Forget it I can't remember all that, being the dumb jock warrior I am, so, I'll just call you Art...ok?"

"Yeah...sure, anyway we'll be stopping at station 16 first. Then along the way you'll get your med check, shots. Then...Triton."

Jarob wasn't sure he liked the shots idea. But, at least he would be away from the Outlaw. Still, it was all happening so quickly he found it a bit unbelievable he would be outta this dump soon. It wasn't until the manacles and tracking bracelet came off, and he boarded the shuttle that the reality set in. Soon, he began to relax, and settle in to the big comfortable chair. Not long after that he fell asleep.

Back home in Oakland, he was twelve years old again, watching a Raiders game with his step-dad, Dwayne. The back of Dwayne's hand came out of nowhere and smacked Jarob hard on the left side of his head. "Get me another brew!" he commanded, as if Jarob should somehow know his beer can was empty. Jarob got him another beer and receiving another slap for his trouble.

His real Dad, Baalor Ka, a Kaldite from the twelfth planet of the Voss system, (a nomadic species that first arrived on earth after their ship crash landed in the desert of Southern California in 2219), had died when Jarob was three. His mom, Jenna, died just last year, some kind of bug or something. When his father died a piece of Jarob went with him. A larger chunk went with his mother, making Jarob feel even emptier inside. Dwayne didn't have any compassion within him so how could he give any to Jarob.

Jarob had typical trait of most Kaldite men, a notch between his eyes, a firm strong chin, and bright, golden eyes as did a strong, compact build, and cat-like quickness.. The softened cheekbones and the general round view of his face, along with very fair skin and curly brown hair were his mothers. He missed her quiet demeanor, her patience, the smell of her hair and that smile he could still see in his dreams.

Dwayne's patience was about even with his compassion. With Jenna gone, Dwayne only saw this half-breed burden he was obligated feed, cloth, shelter, and generally take care of.

Well, he'd do it but he didn't have to like it. Four years of hell followed. At sixteen Jarob was fifty pounds heavier, his own compassion waning he hung with the wrong crowd, getting his education in fighting, drugs and hatred on the mean streets of Detroit. Since Dwayne was little source of money he got his cash by selling Pyresium, N-pac, heroin and anything he could steal. If Dwayne had found out he'd have beaten him for not sharing the profits.

Then, in April just before Easter, his secret came out and

exploded right in Dwayne's face. Jarob hadn't been to school in months. That wasn't unusual for him and he knew Dwayne really didn't care where he went as long as it was away. Usually, he'd fake Dwayne into thinking that's where he was going. Then he'd hang with the guys at the gym for a while, and go back home. Unfortunately, today Dwayne came home early, apparently he went to work a bit drunker than usual and got himself fired, course it was all Jarobs fault.

In his rage Dwayne slapped Jarob very hard, several times. Giving credit where it is due however, it must be said that Jarob was more than patient. Several times he tried to calm Dwayne down. But, anything he said served only to make Dwayne angrier. It wasn't until Dwayne grabbed an empty Tep bottle, and went after Jarob with it, that Jarob began defending himself in earnest. Dwayne broke the bottle on the edge of the a table and looked with eyes full of hatred at Jarob. Wildly, almost blindly, he stabbed at Jarob. Without thinking Jarob blocked, locked, and snapped Dwayne's arm at the elbow.

As Dwayne screamed in pain, Jarob, struck once, hitting Dwayne on the right side of his jaw sending teeth and blood flying in all directions. Again he struck, feeling the ribs snap beneath his fist. Dwayne cried out as best he could. Jarob knew if he remained he would kill him. As Dwayne crumpled to the floor, the empty Tep bottle still in his hands, Jarob grabbed Dwayne's collar, leaned in and whispered "Don't look for me or you die--- asshole!" Roughly releasing him Jarob turned and left slamming the door so hard he heard the glass shatter behind him.

"Come back here you little bastard! I'm gonna call the cops!" Dwayne managed to yell as Jarob left.

Jarob didn't care if Dwayne died or not. All he knew for sure was that he'd finally seen the last of him. Later he had a friend call the cops to see if Dwayne had fulfilled his threat. No warrants, no MPR (Missing Persons Report), no nothing. It figured. Then Jarob really went empty inside, and any compassion that may have remained died.

Time and temper made trouble easier to find. That and a general lack of skills gave Jarob little option. So work was usually running dope, dealing, or generally doing anything for a buck. Another four years later, and he saw a reflection of his younger self in the eyes of another desperate kid. This one held a MAC-50 in his face, commanding him to put the Pyresium in his car. He saw a lot in those eyes, fear mostly, and that same lost misery he had at that age. That same, nobody counts but me attitude, like love had been there only sparingly in his young life as well. It had been so long since Jarob had gotten or given anything like love or compassion, he understood.

This kid his scared brown eyes Jarobs gold eyes staring back, behind them both the same feeling. The kid was unprepared for his attack, and Jarob struck harder and faster than he meant to. He hit the kid, picked up his MAC and without thinking, fired off three rounds, almost as a reflex. The last reflection of life in the kid's eyes, said it all. Jarob had killed a lot of men and women in the arena since then, but that young face and those dead eyes haunted him still. The security patrol saw everything except the kid holding the MAC, grabbed Jarob and well...

"You Ok?" asked Art shaking him awake. "You had a hell of a long nap kind of restless too!"

"Just dreaming" Jarob replied as he looked at the reflection of his eyes in the transparency of the viewport, and remembered.

Computer end recording...

Freedom

Station 16, a remote station at the intersect point between the Sol and Reticuli systems, was one of a few hundred such bases strewn throughout the known systems in this part of the galactic arm. In general it was a safe zone for the scum of the inhabited planets in the quadrant. But, the food was real, good, and kept in constant supply by regular shuttles from Sol, Harin 6 and Zeta Rerticuli systems as well as Berinlos colonies and several other local planets. The strict automated security forces, however, made it impossible to get away with anything major.

FOOD, real honest to God food...steak, mashed potatoes and gravy...could have been gravy, could have been Voldarian swamp sludge, it was so good Jarob didn't care, and peas...real peas not freeze dried and hard like little green pebbles. Even the lumpy brown stuff he couldn't identify was great. He ate like a man who had been starved for the past five years gulping down two large steaks, a mountain of mashed potatoes and a big pile of real peas, topping it off with no less than three strawberry shortcakes.

Then he noticed something else he hadn't seen in a long time WOMEN. The reputation he made for himself among the women of station 16 during the two day layover (if that word isn't too suggestive) would fill another volume of this tome. So, in order to expedite matters suffice it to say if it moved, was female, had a pulse, no matter how weak, and was willing, you can let your imagination go nuts from there...Still, there was this one redhead he saw, but by her look he could see she was in no mood for his foolishness. Jarob, not being the shy one nevertheless sent her a lavender rosebud.

Art, when at the station, spent most of his time in his quarters, feeling that if he were to get in Jarobs way, Jarob may not be able to control himself and in his lust make a grab for him. It was a weird premise, but you never know with some guys, especially those who've been cooped up for such a long time. Actually, Art had spent most of his time off station, busily picking up other recruits, from other systems.

He had just returned from Amna with his last pick up of the week, and was trying to make up for lost sleep, when Jarob knocked on his door.

"Yeah...who is it?"

"S'me Jarob."

"Just a second." said Art kicking the as yet unpacked bags from in front of the door.

"Yeah...what'd you need?" he asked opening the door.

"Nothing...I just thought I'd stop by and see if you're ready to go."

"Yeah...yeah...well no...Actually I...I a...I'm gonna stay here on the station for a few more days.

"Oh...why?"

"Well, I've completed my assignment. You and the others can go on from here. The shuttle is freight and automated so you should make it there in a few hours. Besides, I need some time off." he replied.

"What others?"

"You know the three others I mentioned before we left Outlaw 9."

"Yeah...I think I remember."

"You should have gotten the message...Well anyway, they've been processed and you will all meet at shuttle bay 103 in about 45 minutes, someone will meet you at the camp, and begin your training. So a...good luck, it's been real, Bye." Art said as he edged Jarob back out the door. Jarob stood dumbfounded for a moment or two, scratched his head and walked down the corridor to his compartment.

The shuttle bay was a noisy bustling place, abuzz with activity when he arrived. The launch sites were all connected around a central hub, with each launch tube creating a spoke in the wheel. A tram was used to take passengers and cargo to the main receiving area when docking had been completed.

As Jarob got on the tram, his eyes locked onto hers. She was Amnan, at least that's what he thought, having only seen pictures of them in magazines while in his cell. She was absolutely gorgeous. Though the very light, powder green complexion took him back some. Her long black hair and shapely form, encased in that skin tight, dark jumpsuit, showing off every line and curve, was enough to make him, ready, willing, and able all over again, despite the past couple of days.

There was something about her eyes though they were sparkling, light green, like fine Chinese jade, with a dark green border around the iris. One look in those eyes gave him the feeling not only of sensuality, but a knowledge and intellect that belied her sensual nature. Though he tried very hard not to stare he'd never seen a woman with green tits before, and hers were large, firm, and round, encased in her tight, and very low cut jumpsuit top.

"Hi." he said.

"Gen tro." she replied.

"I...I'm Major Jarob Ka." he said using his new title. (After all it had served him well with the other ladies).

"Pageena Timbo. Warrior of the third gate."

"You a...you on your way to Triton?" He asked.

"Yes...is it your business to know my destination?" She replied.

"No I..."

"Then why ask?"

"Look I was just trying to..."

"To imagine me naked, your pulse rate has jumped, your pupil dilation, and the cloud of pheromones you are putting out all indicate your true intent. I've heard a lot about you Major Jarob Ka." she said with a wry smile.

"Would you be so kind as to let me complete my sentences before you put words in my mouth."

"Ok. What were you going to say."

"Well...I was just trying to make some conversation. It's gonna be a long trip and I'd like someone to talk too."

"Why? Talking is such a waste of time. If you want what you call sex, I can give you that. It's been two seasons since I last joined, in your time that would be six months. But, I warn you do not approach until you're sure you're ready."

"I...eh."

"You better watch how your answer her" a voice chimed in from behind him.

"Benni Jack...Benni Jack Taylor. Like I said you better watch out or that bitch will fuck you into the deck plates. I seen a Amnan girl just like her "join" with a guy for hours once, poor guy still walks with a limp." he said throwing his bags down.

"Oh co'mon you can't be serious...hours." Jarob whispered looking first at Benni then at Pageena.

"That is our way of communication and education, sexuality or at least what you think of as sex, is our life. Amnan males have...shall we say...more stamina than I have been able to find in most other males. Earth males are usually very poor in their performance. Our kind has been known to be in joining for days...Pageena Timbo...gen tro." she replied extending her hand to Benni.

"Gen tro." he replied shaking her hand in a formal greeting.

As they were talking, a solitary figure entered the shuttle, sat down in the corner next to the hatch. The woman was quiet, with a look of calm serenity, and confidence about her. At first no one took notice that she was there, she just sat there quietly, listening.

"I'm Jarob Ka."

"Just call me Ben."

"Where you from Ben?" Jarob asked.

"Mars Colony 6...But for the last six years I was stationed as a CO with Delvinian Security System. Wasn't that fun, how 'bout you?"

"Earth...and Outlaw 9."

"You a CO or...? " He asked

"No...the other side of the force field." Jarob replied.

"Sorry...mind if I ask?" Bennie asked

"Murder."

"Murder...and they let you out? The program must really be getting desperate."

"Not really, I was just the best warrior in that facility. Anyway we all had to kill to get to where we are now."

"Not all of us." Came a voice from the mysterious woman seated by the hatch.

"What?" Jarob asked.

"Oh God.....an OD!" said Ben as he turned to see who was speaking, then shook his head.

"Not all of us had to kill to get here. I'm a Tech 1 they recruited me for my mental superiority."

"Superior thieves." Ben whispered looking at Jarob.

"I am Dafora Meps, Oberon Delta II. My species is superior to every other species we have encountered in the known galactic sector."

"Superior?... superior at stealing the technologies of other systems maybe or superior at profiteering, maybe it's just superior at being the most uptight, stuffy, self-centered assholes in the entire galaxy...hum?" asked Ben.

"Your reaction is typical of most inferiors. Please don't take this the wrong way, but you handle the savage end of this and I'll handle the Tech."

"Don't take it the wrong way? Oh I see there's a right way to be told you're inferior. Excuse me." said Jarob as they turned away and continued to talk among themselves deliberately excluding her.

"I didn't have to kill to get here either." Pageena confessed; "But I'm sure my situation is quite different from yours."

Dafora Meps had been raised with the absolute belief that all beings not of her world were inferior. In fact to gain a visitor's visa to her world you would have to pass an I.Q. test. The planet-wide mean average for I.Q. (as measured by themselves of course) was 565 the human norm of 120 was that of a moron to them. So very few humans were allowed still occasionally they went slumming as they would call it. But normally if you got less than the human norm entry denied.

The entire population is smug, snobbish, and generally the kind of people most of us underlings try to stay as far away from as possible. Dafora's people had a flaw even they would not admit to. They all looked the same.

When I say the same, I mean they all had hair that was white to blond on top and brown to black as it went down the sides. They had high brows and high foreheads, sharp turned up noses that seemed fitting for their attitude, and large purple eyes with yellow flecks. The entire population had an average height of five feet.

The woman and men are almost boringly similar. The only

differences being the women have breasts, usually a bit large for their size. However, according to OD II law the women were to wear restrictive harnesses called Ootman, to take emphasis off their tits and allow for more purely intellectual pursuits.

Many of the other systems thought Deltans to be too smug, and too Victorian in their moral code. In fact most of the general consensus was that the OD II's took themselves far, far, far too seriously. Their primary exports include video games, challenging mental puzzles, I.Q. testing equipment, and (now get this), a mind numbing, intoxicating liqueur called Tep, made from a berry grown only on their home planet.

At first there was some controversy and much debate about the production and sales of Tep among the various elders of OD II. However, since the creation of the Pariki puzzle. A puzzle based on punishment/reward, giving you a minor shock for the wrong answer. It was only after people began to get electrocuted that they discovered that they have used the wrong voltage and had killed people. A simple mistake, they of course denied it and even though they were apologetic, sales plummeted and lawsuits were prolific. After that the request for new puzzles and games dropped dramatically, creating a financial dilemma for the entire planet. The only option that the OD II'S could think of was Tep. (You'd think with all that alleged superior brainpower, the OD's could have come up with thousands of options, but all their stolen technology was already in use and they couldn't think of a way to improve it or otherwise profit from it.)

The seat belt sign came on. Jarob sat next to Pageena, and Benni sat to his left. Dafora stayed where she was. They all buckled in, and a few seconds later the shuttle began its travel down the launch tube. As they cleared the station, the main drive system kicked in and the acceleration pressed everyone into their seats until the dampers came on line.

"Stand by for Hyper-net." a recorded voice blared over the com link.

Hyper net is widely used throughout the known galaxy. Essentially, it is a network of artificial worm holes, produced by creating high intensity magnetic field within a quantum sub-space interface. Once the net is active, it creates a two-way link between the target zone or destination, and the point of origin. A sub-space field of varying flux is then wrapped around any ship entering the net. Usually the field is negative to the nets positive. Once inside the net, the polarity of both the sub-space field and the net are reversed, in effect shooting the ship out the other side. All ships are encoded with different frequencies in the EM band to keep them from crashing into one another.

Most of the time, if two ships try to enter the net either at the same end or on opposite sides, the auto relays place one on standby, until the other has passed through. If not they are spaced apart enough to keep them away from each other. Once through the net, standard impulse ion engines are used to go from one sector to another. Hyper nets were found to be many times more efficient than the old warp systems, far more stable than other artificial wormholes, and in many cases created less ionization of the dark matter, reducing the amount of power necessary for deflectors by almost half. In most cases depending on engine efficiency, the standard warp engines could only achieve hyper drive in the +2 to +4 range anyway. Since no power outputs from the ships were required in the net (aside from the minimal amounts used to drive the sub-space field, deflectors, and life support systems) fuel savings were dramatic.

"We'll be at Triton in 6 hours 56 minutes." the recording again chimed in. "You can feel free to walk about the cabin."

Computer End Recording...

Back to Hell...More Training?

Conversation was sparse during the journey, perhaps it was because of the mix, or maybe it was just that everyone had said what they wanted to say before they left. Jarob sat and watched the distant stars as they whizzed by in a spectral flash. Benni Jack grew antsy and began to squirm around in his seat.

"You know you can get up and walk around." Jarob said.

"Yeah...I know...but go where?"

"We'll be there in a few hours. Just try to relax. Maybe catch a nap."

"I...I never could sleep on shuttles. Think I'll see what they got in the vid-bank." Benni replied as he got up and went to a view screen at the far end of the room.

Jarob's attention once again returned to the stars, and his memories went back to his first battle in the Warrior program. His first opponent Kuutuk, was a nine foot Colarian. The battle lasted almost two hours, with only one minute between rounds. They fought with Sytal, a long curved weapon, with twin blades serrated at either end. The Sytal appeared to get heavier with each passing round. Jarob knew his strength was failing, he also knew how outmatched he was. Even with the handicapper on.

The bout had been arranged by an administrator, who didn't think Jarob got the kind of sentence he should have. As it turned out the Administrator was cousin to the kid Jarob killed (no conflict of interest there.)

The battle continued, on and on until Jarob was sure this giant would sooner or later strike the decisive blow. Jarobs legs grew tired and rubbery, each swing of the Sytal sapped a little more of his strength. His legs failed and he ducked low on the Colarians attack then thrust upward in an act of pure desperation. The Colarian went limp, and sagged to the floor.

When Jarob woke, the Planet Triton was visible out his view port. His heart leaped to his throat as the shuttle closed in on it, and again as the landing cycle relays kick in. As the shuttle landed in bay 4, the drone of the secondary fuel pumping station boomed and rattled the ship so hard Jarob thought they had crash landed.

"OK...you skag rats...get the hell out of there and line up on the blue line." a voice called out as the shuttle doors opened.

At the end of the gangway stood a tall, muscular, black man with long white hair, and bronze eyes. He wore tartan leather armor, with large diamond shaped brass studs protruding from various locations. His huge form leaned on a long black staff of graphite that was highly polished, and engraved in a language unfamiliar to anyone in the group. The long black cigar hanging carelessly out of the right side of his mouth, smoked heavily as he puffed and looked at the shuttle door.

"Co'mon...co'mon hurry your asses up...we already burned to much daylight to do much more than piss and eat today, but we'll more than make up for it tomorrow. Now get the hell outta there so I can see what kinda pussies the assholes at HQ sent for me to get into fighting order!" he yelled.

They could all feel the heat as they exited the shuttle and the pressure made it hard to breathe. Benni was first out, Pageena, Jarob and Dafora followed.

They lined up on the blue line that was directly in front of the shuttle. At first he didn't look at them when he spoke.

"Ten-hut...!"He barked as the team stood with hands in pockets and a dull look in their eyes.

"Holy suffering stigworms!? Don't you assholes know what ten-hut means!? No...no don't tell me...they didn't teach you anything when you were a grade one warrior...Right!?"

"I already been through that shit! We all have...we're grade six warriors now...we ain't gotta do this shit no more." said Benni.

"Oh...oh...oh, I...I...I see, you're all grade SIX warriors now...Oh...I'm sorry...I'm sorry. But....I have one question. Any of you fucker's ever been to Mars II?"

"No...I went by it once." said Dafora.

"In that case... listen up you smarmy pack of lounge lizards. You are in the Warrior Program, which in case you forgot is a military organization so...by it, don't mean shit! Furthermore, around here grade six don't mean shit, and from now on you don't mean shit! So you will do as I say, when I say or, how I say! If you don't we can arrange for you to go back where you came from. Now let's try again TEN-HUT!!"

Immediately they all stood at attention.

"My name is Lucifer Campbell...I am your Kajeem...so you will address me as either Kajeem Campbell or just Kajeem! Now in order you will sound off your name, warrior rank, simulated rank, and the facility from which you were recruited."

"Benni Jack Taylor, Warrior 6, Major, Mars Colony 6."

"Dafora Meps, Warrior 6, Major, Oberon Delta II."

"Pageena Timbo, Warrior 6, Captain, Amna."

"Jarob Ka, Warrior 6, Major, Outlaw 9."

"Ok you pukes, listen and listen hard. For the next ten weeks you are mine, you will eat, sleep, shit, shave, and shower as a team. I want you to know each other inside and out. Now there are some of you who may complain that you are a team consisting of two men and two women, and long dead moral codes or some other bullshit may make you hesitant to live in the same quarters. Just remember this is not a pleasure cruise, nor is it a time to let modesty interfere with your training. If any of you have an aversion to naked flesh, speak up."

Dafora instantly stepped forward.

"Well. What is it?" Campbell asked.

"The superiority of my race, has in past found, that the gratuitous display of flesh, has the effect of..."

"The effect of what...? What's your name kid?"

"Dafora Meps."

"Yeah...well Meps, I'll tell ya something. I don't know many guys who would fuck an ODII on a bet...with another mans dick. So I would say your virginity will remain intact during your stay here. B'sides you and your team are gonna be so dog ass tired at the end of the day even your teeth will sweat. One more thing, if I hear any more of that superior crap come out of your face, I personally will tear you a new asshole. Now get back in line...Anybody else got beef?"

Kajeem Campbell glared at each of them, but more specifically at Pageena.

"I like gratuitous displays of flesh." she smiled.

"Anything else I should know?" he asked.

A round of subtle no's and shrugs then followed.

"Ten-hut!" he again commanded. "Now that we've got the pleasantries out of the way the first order of business is for you to receive your uniforms consisting of your training clothes, dress uniform, and battle clothes similar to my own. Also you will receive bedding and personal hygiene materials, from the quartermaster in pressure dome three also you will inform him of any special needs you may have and he will see too them. From there we will double time it to shack 245 and square it away, preparing for my inspection at 16:00 hours. That's two hours from now. If you pass muster on the inspection, you have orientation and operations lectures until chow at 18:30. One thing more remember always that you are a team, that we are a team, if one of you fucks up, you all pay the price. That's all...ready...on the quarter-turn....double-time...move out!"

In pressure dome six the atmosphere had been adjusted to Mars II normal. The pressure and atmosphere were roughly 1.5 that of Earth, 1.9 that of Amna and 1.3 of Oberon Delta II, so it was a little harder for Pageena to adjust to than any of the others. 05:30 a.m. the next morning they woke to Campbell banging the steel framing of the shack with his staff.

"Ok babies...it's morning...co'mon get outta those bunks. We've got a long day ahead of us...move it out, move it out! Line up in the main quad in ten minutes!"

Now you may wonder what the sleeping arrangements were? First, at Dafora's insistence the men were to remain on one side of the room, the women, on the other. Either side of the shack had a set of bunks, with a footlocker and another tall locker at its head. I explain this for those of you who may not have had the enchanting experience of military service. Dafora also took the position that the women should shower first and the men should be last.

"After all." she said. "We are warriors, not barbarians."

Both positions were unpopular with the others, but it was either agree or listen to Dafora bitch and complain for the rest of the night or give in and try to get some much needed sleep. Unknown to Dafora however, was the close proximity of Campbell's quarters with the shack they occupied. Every complaint, every critique and every little word had been branded in his memory.

The training uniform was little more than a strap that attached to a thong, with a cup to cover genitalia, a harness with a strap for one arm was attached to the thong on which a small cylinder of O2 was attached as a supplement if needed. The uniform (if you could call it that) was the same for both sexes, no special supports for the female anatomy were incorporated. Training in this atmosphere would be hard enough without the added burden of bulky, tight or otherwise restrictive clothing, besides the Women's Coalition of 2542 had broken down all sexual barriers as un-important and sexist in all conditions.

You may wonder further, why in fact the attempt was made to wear anything at all. Basically it was bugs, a particularly viral form of bacilli that had a penchant for genitalia. This bug was initially created to eliminate a disease called Miner's Clap, some kind of V.D. from Zet

Alpha, brought to the facility by a drunken bunch of Tiralian iron hangers during construction. The simulated atmosphere caused a mutation and the bug became lethal. Once a year they would spray for the little pains, but still they have to watch for signs of infection, which I might add are none to pleasant to even think about. But, there has been no sign of any infection for years now, and the warrior program is considering altering the spraying to as needed.

As they lined up on the quad still half asleep and a bit groggy Campbell began his morning sweetness.

"Well it's another great fucking morning...today we're gonna start with a ten K run, then we'll have some chow. Then we'll attempt to make that rats nest you call home, more presentable. After that we'll try you out on weapons training, see if you can keep from blowing your faces off, then maybe run another ten or fifteen K. Everyone inside and suit up in training uniforms."

"Ain't you afraid your tits are gonna fall off, with all the running." Jarob said to Pageena as they were dressing.

"Their not yours, so I wouldn't worry about it. Besides, your gonna be too tired to care anyway...remember."

"It'll never happen." Jarob said, halting and retracting his statement in an instant, the moment he realized who he was talking to.

Dafora was wearing her Ootman, with her training uniform over it. After they had finished dressing (or maybe undressing is more accurate) they again took positions on the blue line in the quad. Campbell looked them over. They all looked ready for the hard day ahead...except for Dafora. Campbell knew that if she continued to wear the Ootman in this pressure with the kind of exertion they would be going through, one of two things would happen. First her heart might just explode, that was the quick, almost painless option. The other being, that over time, if her restrictive harness was not removed, she would develop something akin to the bends. Either way she would be useless to the team. If he ordered her to remove it, her arrogance would tie the order to some bullshit sexual thing and he'd get flack for that for the next ten weeks. So he'd run them all, hard, if she died, he'd get a replacement, but bullshit or not the minute it became a problem, he'd remove it from her himself. All this information went through his mind as he stood face to face with her.

"Warrior." he bellowed as he looked into her eyes. "The Ootman you wear may be proper and necessary in your home world, but, it is ill advised here, not to mention the fact that it is not regular warrior issue. Now it is not my place to order you to remove it. Be advised that I have told you, in front of witnesses, the minute I see that thing slow you down even a little bit, I swear to God I'll rip it off you personally, regardless of your modesty or moral bullshit...You got that warrior!"

"Yes...but..." she replied.

"But what but you wanna die...you wanna let down the team? Your precious tits are worth more than our lives worth more than your own life? I don't think so...I don't think they think so either. So you got two choices either ditch the Ootman of your own free will, or prepare to spend a very embarrassing day...and believe me in this pressure and atmosphere you'll be lucky if that time doesn't come within the first couple o' K...Clear!"

"Yes, I see Kajeem." she replied.

Dafora didn't remove her Ootman, her arrogance just wouldn't let her admit the fact that this inferior creature, could possibly be right. She did ok for the first five K, during the second, she dropped like a stone. Campbell motioned for the others to continue and knelt down beside her. He took her pulse, it was racing but solid, and true to his word he did exactly what he said he would do. The Ootman was history, which proved to be a bit more of a chore than he had originally intended it to be. There were fasteners up the sides, down the front, and down the back, which all interlocked. Finally he just said "Fuck it." and cut it off.

"Damn." he said "These babes must have tits of gold." He muttered as he sliced through the tough material. Eventually, the task was completed though she was very slim and probably just under 5 feet tall, her small frame made her boobs look mountainous. "Holy

God." Campbell muttered trying not to notice as he slapped her lightly on the cheek to revive her. That didn't work so he tried a little mouth to mouth.

In a start Dafora woke up, and looked down at her exposed breasts, then a Campbell, shrieked, cupped them, and began to run back toward shack 245.

"Just where the hell do you think you're going!" He yelled.

"I...I can't let the inferiors see me like this."

"There's no one here who is inferior, and there's no one here who give a shit about your tits! Now, you get your ass back here or you'll be on the next shuttle out!"

"But...but I've never shown myself to anyone...not anyone." she said as she moved back toward him, with her hands tightly covering her breasts.

"Honestly, I've never understood the Oberon Delta smug morality, or their reasoning behind their superiority complex. Can I ask you something? Why? Why in the hell did you apply for the Warrior Program? I mean others of your kind have been in the program I've trained 'em, and we've never had this trouble before?"

"They were all men, I am the first woman of my species to apply and be accepted."

"Why? I mean didn't you know about the training harnesses and the conditions here before you got here? I can understand with the others. But you, you're a tech, what are you trying to prove anyway?"

"Prove? I...I'm not trying to prove a thing." she lied.

"You're not huh? Ok have it your way for now...But, if your ODII bullshit gets in my way again, you'll be out on your ass. Now shag it and get up there with the others."

"But...my breas..."

"Are not as special as you think they are...now you can go with the others or get the fuck home...little girl!"

Dafora began running toward the others with her hands still tightly clutching her breasts. Campbell only shook his head and followed. Dafora had to admit however, that the run without the Ootman was both more comfortable and allowed her to breathe more easily. For half an hour she managed to cling tightly to her precious bosom, until at last her arms could endure no more and dropped to her side. Jarob and Benni were about to make some comment, when Campbell piped up in a booming voice.

"Any of you assholes, opens his big, fat, mouth and you'll get three laps around the dome!"

The dome was three kilometers in diameter, and since no one was ready to run an extra nine kilometers, tongues were silent. Especially since they had just completed the ten K and almost died from that. Wheezing and puffing Jarob rested his hands on his knees at the end of the run. Pageena landed roughly on the ground beside him, sounding like someone suffering a severe asthma attack and taking a large sniff from her O2. Benni and Dafora suffered similarly when they too had finished.

"You slack-jaws think this is rough? From now on you'll get two runs a day just like that one. Except for the last two weeks...then you'll run in full battle gear." Campbell yelled. "Now, I hate to say this but I've got some bad news. After chow we won't be doing a barracks inspection after all. Kajeen Ma Coorbas, the director of this facility, will hold an orientation lecture and briefing. Since no one knows how long that will take, you kiddies may miss out on your afternoon run, but don't be too disappointed, what we lose today we make up tomorrow. Go get some chow, then shower and change into your dress uniforms, and report to building C at 12:45. Let's get to it people."

No one felt too much like eating, sleep was a different story, but not eating. Still, they managed to choke down some powdered eggs, half raw bacon, and a cup of coffee or two...or five. When they got back to barracks, Dafora started in again with the women shower first shit, which started another bout of who does what and who sweats the most. Campbell came in about the time it had almost come to blows.

"What the fuck is going on here!"

Immediately everyone began to speak at once.

"Dafora...is telling us guys that we should shower last cause..." said Benni.

"I'm full up to here with her bullshi.." said Pageena.

"Either get her out of here or...." Jarob yelled.

"That's enough...that's enough...THAT'S ENOUGH." Campbell yelled above the din. "What's the big deal it's a four stall shower, you can all shower at the same time! Dafora, you will do well to understand that the only one around here who gives orders is ME!! Now you two guys go on and shower, I got to talk to Pageena and Dafora alone a minute."

Campbell took Pageena and Dafora aside.

"Now I don't know if you know this, but you two are not the first women we've had here. We've been training this way for centuries. Whether you like it or not this is your team, any more shit from anyone about the difference between male and female and this team will be disbanded and reassigned, is that clear?"

"Yes Kajeem." they said in unison

"Just in case you're wonderin' I already told the guys earlier, I didn't tell you together because there's something of a more personal nature I wanted to talk with you two about."

"Oh?" said Pageena with a smile.

"Wipe that grin of your face woman!"

"Yes Kajeem."

"Now, in the past the training uniforms have been structured to reduce physical stress, and enhance breathing capacity by being non-restrictive in nature. However, the amount of running, jumping, and other exercises, has proven to be hard on women who are, shall we say, well endowed."

"You mean women with big tits?" asked Pageena.

"To put it bluntly...yes. New training uniforms for such women are on order. But, they haven't come in yet. So, I went over to the PX and got this black frilly thing and this pinkish deal here. Now I...I don't know what they're called they just got these cup dealies and I...I don't know nothing about sizes and shit, I just got the biggest one's they had, you wear 'em if you want to. If you can wear them as loose a possible around the ribs that would help. But, I warn you if they get in the way...I'll cut them off like I did the Ootman and you won't get them back...you got that!"

"Uh...Kajeem, we don't have to wear these...these things, do we? asked Pageena.

"No...like I said it's your choice."

"No thanks, I'm fine." said Pageena.

Dafora on the other hand, lit up, grabbed the bras and smiled so big it was hard to tell that this was the same woman who had been such a bitch only moments ago. Campbell turned to leave, as he did he and Pageena locked eyes. Both knew there were no new uniforms nor would their be any forthcoming.

"Dammit we're a team." he whispered to Pageena. "If this'll make thing work more smoothly...."

"You're pretty nice, for a son-of-a-bitch." Pageena whispered back.

"OK people...you got twenty!" Campbell yelled as he slammed the door behind him.

"So aah...what was that all about." asked Jarob as he exited the shower, drying his arms.

"Girl talk." she said going in.

"Girl talk? With Campbell?" Jarob said looking at Ben.

Computer End Recording...

Meet the Enemy

They entered building C and took seats behind a large wooden table, on which sat a pitcher of water, glasses and four briefing booklets, each inscribed with their names. The lights began to dim and a large monitor screen in front of them came on, before them on the screen stood another man in a blue pinstriped suit, with a large smile on his face.

"Hello, and welcome to the Warrior Training Program, I am project Director Coorbas. Though I cannot be there personally, please be assured that I am with you in spirit. The training program you are now undergoing will be tough, hard, and brutal. Once completed however, you will be placed on a rotation list for service on Mars II. More information about Mars II will be given at a later briefing. For now suffice it to say, that the atmosphere and pressure in which you now train, will be beneficial to you there.

Your team will consist of four members plus neutral, including a Tech, a tactical, and three warriors. The job of tech is to read enemy defenses, collate the data and give the information to tactical, in order to coordinate the necessary attack profile. The warriors are far more than just fighters they are the backbone of the team. You are all killers, but to do that the warriors must have the right information at the right time, and be able to understand and carry out instructions.

A successful team must have a good and effective line of communication. There can be no power plays, no angers that create entanglements in communication, or one person within the team who has more or less power than any of the others. You must be unified and based upon an absolute goal; that goal should be the total destruction of all enemies. It is not certain, but likely you will face at least one of the teams in this briefing in battle. The most likely will be this one."

The image of a large, hairless humanoid, with a prominent brow, and deeply inset fire red eyes, stared back at them.

"The creature you are looking at is an enemy. They call themselves the Coonwaadii, they are a warrior race of high intelligence, level 6 technology, and have been beaten only once, in 2914 by the Impet Cassoniga on Mars II. The defeat was by default, so it would be fair to say they have never been beaten. In 2927 they annexed an area three parsecs wide and 5 deep, containing nine of the most productive planets and mining operations in the quadrant. These areas have been in dispute since that time and the Coonwwaadii have shown no propensity for negotiations, nor have they been open to any refugee release programs.

As part of this briefing, the information in the booklet before you must be learned and committed to memory. Coowaadii battle tactics are usually flank, sweep, and drive. They are merciless in their attack, as well as in their treatment of prisoners. But the rules of war (which will be covered at a later briefing) state that humanitarian factors are not an issue.

The Coonwaadii are led by the brothers Bosrum and Sokeiva Pocto. It is usual for them to work together with the rest of a five man team. Their success we believe is the fact that they have had the same team for many years. So long in fact, we believe a family relationship has developed. A united family as you all know is very hard to beat, if for no other reason than communications are tight and they are well organized.

Bosrum works tactical, Sokeiva usually takes the position of Tech. Others in the team include, warrior Jimba Ti, who is a 9'2" Dorelian with a weight of almost 1300 lbs. and is their point man. For spotter and sweep they have warrior Tinnic Ba, a Zeluthian. Well known for their ability to see in a light spectrum far beyond most humans. Finally, warrior Uberus the Porceinan, she fights hard and is another power player with skills in a great many types of weapons.

Get to know these opponents, get to know them all, very well. The key to your success may well rest in your ability to find a weakness in there team and exploit it. Included in your briefing booklets, you will find detailed information on each of them. It is not a certainty that your team will be the one to face the Coonwaadii. For that reason we have kept more pertinent information out of this briefing. But it is standard practice that all warriors receive some information to keep them up on operations in the works.

In about eight weeks time, a team will be sent to Mars II to face the Coonwaadii. We tried to get an earlier time in the interest of expediency, but the referees have said that nothing can be done until the Kotarian conflict has ended. After that Mars II is slated for the Ypzlatizian's. So, we have at least that amount of time to get a crack team together.

There are twelve lectures in this series, you will have one each week and two the last week of training. The booklet is your bible, do not lose it. As you can see each booklet has your name and position written on it. This is so it will be tailored to your specific part of the operation no one can use it but you.

Special equipment will be issued to each of you at the end of this lecture. If you are a Tech you will then proceed to building 2, level nine. If you are a Tac you will proceed to building 4, level 1. If you are warriors you will go to your Kajeem, who will then instruct you in use of the equipment. A Tech is pre-determined from your team, however if you look at your booklet and the letter T is in the top left corner, you are the Tactical designate for your team. Good luck and Thanks again."

The lights came up and instantly everyone, except Dafora looked at their briefing books. Jarob was happy he hadn't found a T on his. Pageena on the other hand was pissed, because she had. As they got up to leave Campbell came in with two battle cases. Each case had a different name on it. As he passed them out, he actually smiled, not a kindly smile, but a devious one.

The warrior's cases were long and bulky, each containing nine items. A plasma field pulse rifle, two copchoc nygor (a weapon worn on the wrist that fires small razor sharp disks about the size of a penny at 1400 fps), a dieapoit (similar to a saber, only shorter and serrated on both sides), a kizanti shield, three Mobor and a standard blaster. The kizanti shield is an electro-magnetic field intensifier, highly personalized and attenuated to the EMF of the owner, effective against most weapons, however the power drain is enormous, and since it draws power from its user, it is best to use it sparingly.

Mobar, are essentially blinding flares made of magnasite. With an output of 3500 photons per cubic centimeter, they can produce a daylight effect of an area of one square mile. Needless to say at close range it would produce blindness, unless protective visors or glasses were worn.

A small pocket in the lower right corner of the case contained a K-16 capsule, to be used in the event of capture. It was both a weapon and a means of self-destruction. It was attached to the weapons belt just behind the buckle; once activated by unbuckling the belt without disarming it would explode within 15 seconds killing everything in a 5 meter radius. To remove the belt without activation required that two buttons on the top of the buckle and one on the bottom be pressed simultaneously. To activate quickly a quick release was built into the belt. Once the capsule was in place any attempt to remove it would activate it. It is the only type of explosive allowed on Mars II.

In addition to the weapons and the K-16, a helmet incorporating a universal translator (in addition to the one everyone had implanted at birth it covered a greater database and a much larger range of languages. Most implants had a range limited to the system one was born in), communications transceiver, heads up display laser-sight, and vid-com downlink camera were included in the pack. In the bottom lay the body armor to be attached to the battle clothes they had been issued earlier. It was light weight and could absorb, and dissipate a standard plasma field charge in the fifty watt range. Up to twenty rounds could be dissipated as well as most high velocity projectiles at close range. However, if you found yourself in a situation where you were taking that kind of abuse, the Kizante shield would be S.O.P.

"To begin, you will all place the K-16 into your battle buckle and make sure it is secure. Then remove all the remaining equipment and place it into the assigned slots in the battle packs at the bottom of the case. Then pick up the pack and snap it into place on the back of your battle armor. Once it is in place, stand by for inspection, and await my approval." Campbell ordered.

After all the gear had been placed in the assigned slots Campbell came over, looked over the packs, told Jarob to adjust his Mobar position or it would activate the moment he grabbed it.

"These wrist monitor's…" Campbell continued giving each of them a small band that resembled an old style watch. "Will give you an inventory at a glance incorporated into the monitors, are a motion detector and tracker. The tracker will show each member of your team using a different pair of letters JK for Jarob Ka PT for Pageena Timbo and so on. The tracker and motion detector can be used in combination so you won't confuse your teammate for an enemy. Finally the detector will only register movement of lifeforms who are carrying weapons. Don't ask me how it works or how it knows what it knows, I'm not a tech, neither are you guys, so just go with what you got."

Dafora strolled into the tech center in building two, level nine. She was met at the door by a tall, thin man with a pencil tucked behind his left ear and a look in his eyes that was a bit frightening.

"I ain't had a woman in eight years." he admitted staring directly at her breasts as if that's all there were to her.

"Down boy…" came a voice from behind him.

He turned and looked at the kindly face of the aging man with white hair and whiskers.

"I'm Doctor Davris Keefru, this is my assistant Dalfeen." He said pointing to the young woman at his left. "And that irritable creature is Marvin Raoul. Marvin tend to the assignment I just gave you."

"Dafora Meps…I was told to report here."

"This is tech center 1 here you will be issued your battle equipment as well as the specialized material you will need to complete your missions. Have you ever field-stripped and repaired a multiphased sensor array? Or perhaps reworked a nictron baridium power exchange chamber?"

"No…I don't even know what those things are."

"This is a multiphase sensor array." He replied handing her a small rectangular box about three inches long and two wide.

"What does it do?" she asked.

"Everything...it calculates incoming assault patterns, tells you what kind of weapons are being used. How best to counter them and what best tactical counterattack to use to activate just press that green button there."

Dafora pressed the green button on the top.

"Phase please?" it asked.

"What?"

"Unless phase is given within 22 seconds this unit will self destruct with a plasma charge equal to five terawatts output.

"Phase gamma gamma alpha 6." Dr. Keefru interrupted. "That is a security measure installed into the system to keep the enemy from tampering with it. Unless the correct phase is given to the voice con activation, the unit, which also has a vocal recognition pattern analyzer, will overload causing an electric discharge powerful enough to fry whoever is holding or even within five meters of it."

"What else does it do?" Dafora asked.

"It can do many things. But, first of all you have to look at this." Dr. Keefru said tossing her a small, thick, brown book. "This is your code book. There are 680 codes in there and you must memorize all of them and there functions as pertains to this unit along with the corresponding backup emergency codes."

"This could take months." Dafora replied.

"Maybe...but you don't have that long. You've got a week. At which time this book will be returned to me. If it were to fall into enemy hands well..."

"A week! One lousy week! That's all! But, you said yourself there are 680 codes here, plus the...let's see...235 backup emergency codes. That's 915 codes...in one week. That's impossible."

"That may be, but other techs have done it in less time. Human techs, Amnan techs, Pholarian tec...." Dr. Keefru said.

"Alright I get the picture." Dafora replied.

"I'll give you a tip or two though. Many of the codes are redundant codes that do the same things in slightly different ways. Some are null codes used when shutting the system down, and some are just plain useless. It is your job however, to get to know this machine better than you know yourself. I will expect you to pass all requirements within one week from today. If you don't you cannot be issued this unit. Now here is an operation's manual a programmed spec outline and your code book...don't loose anything...I'll see you back here in a week." He concluded.

Dr. Keefru then turned and walked away leaving Dafora standing there. As she turned to leave Marvin was once again right their starring in awed fascination at her chest.

"Get out of my face...fool!" she yelled as she stormed out.

Pageena on the other hand was having problems of her own. Upon arriving at her duty station she was given a Tactical Plot Analyzer (TPA for short) and given three minutes to give coordinates, speed and distance of a weapons bot at the far corner of the compound. Though she knew how to use the TPA, and could under normal conditions, have completed the task, the conditions were not normal. On her head they placed a set of earphones with the sounds of a pitched battle being fed into them. The low lighting was interrupted occasionally by high intensity strobe lights, and blasts of hot air. What should have taken three minutes, took an hour.

"Not good enough! Do it again!" the instructor commanded over and over again.

Finally after five hours of this hell with the sounds of bombs throbbing in her skull she managed to cut it down to twelve minutes.

"Twelve minutes...twelv!...You think you got it down with twelve minutes! No...no I don't think so...I think you should do it again!" the instructor yelled.

"I got a goddamn headache." said Pageena as she walked out of the room.

"You be back here tomorrow by 09:30 or your ass will be on its way home! You hear that Timbo...09:30!" the instructors voice faded as she walked toward the dispensary.

It was now 18:30, chow time. But she wasn't hungry neither were Dafora, Jarob or Ben. The meat looked old, the bread had mold and the potatoes were swimming in a thick greenish ooze they called "gravy."

"This sure ain't station 16." said Pageena as she popped another hotracine table to try and quell the headache and washed it down with some coffee, or at least what pretended to be coffee.

"You got that right." Jarob agreed.

"Man. I packed and unpacked my battle pack so many times today..." Ben said.

"I know...I was there remember." Jarob replied tersely. "Just be glad you two don't have battle packs to carry around...those suckers are heavy." he continued looking at Dafora and Pagenna.

"They do." came the deep strong voice of Campbell. "They have the same pack as you do...minus an item or two because they have other things to carry. But essentially they all weigh the same. Ok boys and girls...tomorrow will get here whether you want it to or not. So I'd suggest you go back to the shack and catch a few z's. We want to be fresh as a daisy and ready for anything on tomorrows first 10 k right? Hustle it up...move it out."

Computer End Recording...

The Ripper...

"05:30...Wake up you bed bugs!" Campbell commanded as he began banging on a trash can with his cane. "Get into your training gear and be on the blue line in three minutes...shag it boys and girls this ain't no day camp!"

Dafora was the first one out, in her new bra and training gear she looked more than a little ridiculous, but she was happy and that made the rest of them feel less like choking her to death. Pageena still had a headache, Jarob looked like he used a rake on his face and Benni was barely sure he was alive.

"We got no lectures so it's gonna be a full day. First a ten K then chow, then hand to hand battle tactics, then course one and then the ripper...oh you all are just gonna love the ripper. Now I know the tech and tach have assignments that will require them to not participate in some of our regular activities. This does not however excuse them from any of their regular training routine. Pageena you have a session at tach at 09:30, when you have finished that session you will report to me at the ripper. Dafora yo..."

"Excuse me Kajeem...where is the ripper!" Pageena interrupted.

"Just ask anyone, they all know what and where it is." he replied.

"Dafora you have studies so for the next week you are excused from the afternoon ten K. But, in the morning your ass is mine. Now let's pace it out you maggots...ten-hut...right face...quicktime...move out!"

Ten K was a little easier as they adapted to the heat and pressure. Running around an oval track though was easier than running the same distance over rough terrain, as they would soon find out.

"OK you pukes...that is the last time you will run this track." Campbell yelled. "You see that small range of hills being built in the center of the track? You boys and girls are gonna be the lucky ones to use it first when construction is completed tomorrow. It will more properly simulate the true conditions of Mars II. There will be trees, boulders, small stones, brooks and even sand to get in your boots, when you wear them. Sounds fun huh?"

"Yeah...loads." Pageena replied.

"Go...gonna be...great." Jarob panted.

"Good...good I knew you kids would like the idea. Now go get showered and chow and meet me on the blue quad by dome five in an hour. Bring you battle gear." Campbell barked.

The showers were cold and soothing as Jarob stood for what seemed to be the entire hour beneath them. The showers were on a timer however, and each allotment only lasted 3 minutes and could only be used twice a day. After choking down more powdered eggs, moldy toast and coffee strong and thick enough to paint with. The team made there way to the blue quad where Campbell and a person they had not seen before met them.

"This is Do-Gen-Kao, our hand to hand instructor. She has successfully completed six tours of duty with this facility and has been to Mars II nine times. She is among the most highly honored and decorated warriors ever. Listen up people!" said Campbell.

Do-gen stepped forward. She was very small, looking as though a feather would kill her in an instant.

"What do you see when you look at me..?" she asked looking at each of them in turn. "A frail flower, perhaps an easy target?" She continued as she glared into Jarob's eyes.

"I....I...Ahhh." was all he could say.

"Come forward." she commanding still looking into Jarobs eyes. Jarob stepped forward.

"Attack." she commanded.

Jarob raised his arms and grabbed her roughly by the throat. Reaching upward between his arms she took a firm hold on each of his thumbs and pull out and downward. In seconds he screamed in pain and fell to his knees. Though she had released one thumb, control of the other was hers. A twist and kick combination laid him on his back, and a nerve pinch in the right temporal region made him cry out briefly, before he passed out.

"That was a clumsy, stupid and easily defended attack." Jarob heard her say as he woke and shook his head trying to clear it.

"Your attack must always be swift, sure and without hesitation. It is always better and more to your advantage to go for a target that is both unexpected by your enemy, and harder for them to defend." she continued as she held out a hand and helped him to his feet.

"The obvious target is not only easy to defend, because there are more moves designed to counter and protect them. Because they are the ones more warriors go to first to try to achieve a quick kill. However, once the quick kill attack has been countered, if no other strategy has been thought through, and only a fool would battle with one strategy, the quick kill may be your opponent's. So, during this training session and for the upcoming weeks we will be concentrating on strategies; targets that are not so well protected and not usually thought to be fatal."

At that point a large Asian man with muscles out to here came in and stood behind her.

"This is Zuo-Sho my assistant, husband and all round good guy. Each day before we begin we will do forty-five minutes of strength and flexibility work. Training will begin after that and last another two hours. You will be expected to learn and perform each of the techniques we will show you as if you were born to them." She continued.

"You two are excused to attended classes, but you will be required to return later in the day. Your Kajeem will tell you when." She said looking at Dafora and Pageena.

"As for you two, limber them up Zuo!" she commanded as she walked to the side of the mat sat down and quietly watched.

Pageena and Dafora left and Zuo looked at the two remaining recruits with a small yet frightening grin on his face. The limbering up was rough, the training thereafter was even worse. However, what was yet to come would truly make each warrior question the real reasons they had entered the program in the first place. As if they hadn't done that already.

After chow they were given half an hour to rest. Then instructed to suit up in training gear, and go to the infamous Ripper at yellow section dome 16. There they were met by Campbell who had a frightening grin on his face. As they approached, they noticed something else unusual. Campbell was staring at the bare breasts of Pageena, something he usually never did.

"What?" asked Pageena.

"I...I'm sorry I should have told you to wear your battle armor over your training gear, at least the top, in your case anyway." Campbell explained.

"Why?" she asked.

"Well...the particulars are unimportant...ahem...suffice it to say that when the ripper was first used by a female recruit a few seasons ago...well...she...ah...she, got hurt real bad...in...in that area." He said pointing and shaking his finger at Pageena's chest.

"What happened?" she asked.

"Well...I...I don't wanna tell you it...it's a bit embarassing." He protested.

"Campbell...what the hell happened?" She pressed.

"Ok...ok. There are a couple of conduits you have to climb up that go next to the main power regulators for the holographic imaging system. The conduits weren't exact designed with women in mind. The sheet lasers run just a few centimeters on the other side of the ladder. Sheet lasers are not dangerous unless they come in contact with bare flesh. This warrior was going up the ladder and well...happened to lean in a bit too far and..."

"The sheet laser fried her tits!?" Pageena yelled.

"Well...well ye...yes and n...no. It...it cut them off first...then frie...fried them. And hers weren't half as big or nice as yours." Campbell replied.

"Gee ah thanks, I think...You got another one of those bra's handy Campbell. I don't wanna lose two of my favorite assets." She asked smugly.

"No...You don't want to lose your assets." said Jarob with his hand on Benni's shoulder, both men straining to hold back their laughter though they knew it wasn't funny.

"Ha...ha. Real funny assholes!" She yelled.

"Alright...alright that's enough of that shit. Pageena hustle back to the shack and get your battle top. You two assholes shut the fuck up and stand at attention until Pageena gets back."

"Uh..Kajeem...?" Dafora asked meekly.

"What!"

"Will I be ok in this?" she asked pointing to her black lacy bra.

"Yeah...yeah I think so. It...it only works with bare flesh. You...you should be fine." he replied

Jarob and Benni stood at attention and giggled to themselves when Pageena returned with her battle top on. They knew it wasn't really funny, they also knew if they didn't watch themselves Pageena could turn them both into eunuchs in a heartbeat. So, it was a mixture of nervous and joking laughter.

"Before we go in you must understand two things. One: most of the things you will see in this room are not real. However, in terms of the danger they represent an actual life threatening situation they are all too real. Two: There are fail safes and protocol programs used to keep anyone from being actually killed, but you can be hurt sometimes quite severely if you don't stay on your toes. Welcome to the ripper."

Behind the two massive steel doors were twelve separate rooms each containing holographic images of various planets in various quadrants of the galaxy. Seven had been assigned a Mars II program covering every possible terrain and simulated life form.

"In this step of your training you will be on level 2 that is only the two rooms in front of us and the two above those. Run simulation program Mars II level 2."

As the door opened the sound of chirping birds and other creatures of the tropical region selected as level 2 began their litany. It appeared so real that for a moment they were caught off guard. When the Nesruu attacked, Benni was killed in simulated attack almost immediately. Loud klaxons went off and he was held in a force field until the system went into auto reset at Campbell's command. The rest of the team got their collective asses kicked badly.

"That was the worst goddamn example of fighting I've ever seen. It...it...it sucked. I mean it really sucked baaad... in fact I can't believe how badly it sucked. Nor can I even put into words the magnitude of how badly it sucked. Now get your asses back on the line outside and be ready this time." he continued.

The system had to be reset four additional times in the two hours they were there each failure causing Campbell to get more and more angry. Pageena was killed once, Benni again and then Jarob. The only one who went unscathed was Dafora.

"Alright!" barked Campbell. "This is the last time I'm gonna reset this damn thing. These enemies are the weakest in the program. If you can't beat them you're worse than worthless."

Finally they managed to defeat the Nesruu without any permanent damage. But, Campbell made them run through the simulation another six times to make sure that victory wasn't a fluke.

"Alright grab your shit and shag it back to the shack!" Campbell commanded. One more ten K and we're through for the day babies. Pageena, you take your battle top back to the shack and don't forget to bring it with you when you come back here tomorrow. All of you are to meet me on the blue line in front of the shack in three minutes. Now move it out you worthless bags...   Computer End Recording...Buffer filled

BEGIN TRANSMISSION

TRANSMISSION COMPLETED

A Step Back

As a junior administrator to the Mars II facility it was Marsha Tate's job to attend the Council scheduling meetings. She hated them, they were boring, and many of the representatives had less that adequate personal hygiene, or to put it less diplomatically they smelled big time. As she entered the board room, it soon became clear that this would not be a usual scheduling meeting. There was a great deal of buzzing and milling going on, making the scene look more like a cocktail party than a scheduling session. At the other end of the long oak table, chair-person Karen Turbin banged her gavel several times and tried on many occasions to call the meeting to order. This had the effect of creating more background noise and heightening the already tense situation. Finally after practically screaming herself hoarse and splintering the gavel, the room began to quiet.

"We cannot allow this kind of disorder in the board room. I understand the situation is not an easy one, but rumors and speculation will not get to the fact. So we will table discussion of these rumors until our intelligence agents give us more information." She said calmly.

"I agree with ambassador Turbin completely." The Xanthian delegate chimed in.

"Thank you. Ambassador Trouque." she answered with an almost sickly sweet smile.

Another rumor going around was about the clandestine activities of Ms. Turbin and Ambassador Trouque. About their naked, hot sex fest about three weeks ago. They were seen completely naked chasing each other through the halls at the Ambassador Suite Hotel at 4:00 am. When asked or questioned about it, they both vehemently deny anything improper or any wrongdoing. Still, the only witness was a bellboy and whose word would the Council take in that case? So, the rumors persist.

As the Council and staff began their meeting, it was decided that Mars II was should be made available for a battle between Earth and the Coonwaadii in twelve weeks. A dispute had its beginning as a boundary issue involving 15 square parsecs of space, and 9 planets recently colonized by Earth. A few months after those colonies were established the Coonwaadii arrived and began annexing planets and their inhabitants. Rumors of Coonwaadii atrocities were publicized throughout the known sectors, and were not really taken seriously until an Earth cruiser, the Dallas, was fired upon as it was about to enter the atmosphere of Delphos III, the ninth planet in sector 297.31.

An Earth Ambassador, all his aids, and the entire crew complement died of suffocation and exposure when their life support systems went out after the attack. The ship bounced off the atmosphere out of control into deep space and wasn't found for three months.

There had also been stories of murders, rapes, and bloody mutilations at Coonwaadii occupation hands. Had the Ambassador not been David Coorbas the Director of the Earth Warrior program, it may have been diplomatically seen as a mere mistake on the part of the Coonwaadii. However, it could clearly be heard in the recording of the transmissions that David announced his presence, stated that they had an emergency situation on board and the Coonwaadii fired on them anyway saying they thought the Earth ship was making an attack run, though full sensor readouts would have confirmed that the Dallas had no weapons.

With that evidence in hand it was considered to be more than an adequate reason to go to war. The Council and most of the Administrators of the Warrior Program agreed, and had tried for a emergency timetable for aggressions to begin, after all people were being killed. But there were other matters of equal urgency the Council had to consider as well. When the scheduling meeting had ended, Marsha left the building and went across the street to the Baltone Bar and Grill. After ordering some steamers, two fingers of Tep, and a beer back, she settled back into her chair just about to relax when Karen came in.

"Hi Mar." she said sweetly.

" 'lo Karen. Just taking a break, those meetings fry my brains."

"No problem. There's just something I gotta tell you about on a kind of informal basis." She said lighting a short black twisted cigar.

"We got a special project. Real hush hush. I...I was wondering if you might consider..."

"Field work, are you nuts. I've been their and done that, I'm a desk jockey now and like being a desk jockey. Look you got agents coming out the ears, ask one of them."

"Can't." Karen whispered.

"Why?"

"Well we can't have any involvement in this. The Warrior Program I mean. Look I can't really talk here, come to my office when you finish your break."

"Look I'm telling ya, I'm really not interes...ted." She said as Karen got up and left in a cloud of smoke.

Marsha finished her steamers, Tep, and beer, and sat at the table in a stalling tactic as long as she could, dreading the "talk" she was to have with Karen. She also knew that she shouldn't have had the Tep, especially with beer, it kind of heightened Tep's effect. An effect that did not in any way need heightening. By the time she got to Karen's office she was in no mood for debate, questioning or even talking for that matter. Luckily when she got there, Karen had been called into an emergency session, Ambassadors only. So, Marsha said screw it and went home.

Home was a two room apartment on the third floor of an old brownstone on the east side of Trenton New Jersey. It was a dump the neighbors were loud drunk and usually blasting Roopa music till all hours. Roopa music was a mix of jazz, heavy metal, and east Indian sitar classics, started about six years ago, Marsha hated it. In fact the only thing she even remotely liked about the place was its close proximity to work.

As usual she was met at the front door by the manager, a large fat man with a gut out to here and the breath of a zegalian wharf-rat.

"Hi girlie." he slurred.

"'Lo George." she muttered as she passed him quickly.

"Hey honey your rent's due soon, you know I'm always available to discuss an alternate payment." he yelled as she walked upstairs.

"In your fucking dreams." she yelled back.

"Yeah...yeah well fuck you too! Ya lezzie bitch!" she heard him yell back as she started to ascend the second course of stairs.

A drunk in the hallway on the third floor had just puked his guts out not ten feet from her door.

"Jeez...I gotta get out of this fucking dump." she muttered as she stepped over the mess and punched the codes on her lock. The Tep had screwed up her coordination some so she had to punch the code a couple of times before the lock opened.

The smell of the puke in the hallway made her feel like she would get sick herself if she had to stay out there much longer. When the door opened she quickly went inside and hit the control for closing and sealing the door. She then grabbed the phone and called George to complain about the puke, even though she knew it wouldn't do any good.

"Hey! George! Some asshole just puked in the hall by my door could you send up a cleaner to take care of it."

"You want it clean? Clean it yourself!"

"Ok...I can do that. But...Mr. Fairfax the owner...and a personal friend of mine...would have to hear about it and I would also have to call the Housing Commission and tell them of the conditions here...about how the water stinks, how the heat work great in the summer, and how the rat population has started it's own brothel in the basement. Do you think they might be a little bit upset? Maybe upset enough to ace your fat ass out of your rent free apartment and find someone who can really do the job. But, this would only be the seventh complaint this month. Do you think there might be a reason to worry?"

"Yeah...yeah, ok...someone'll be up in a while."

"Make it soon fuckhead!" she yelled slamming the phone down.

"What an asshole!" she yelled as she threw her purse and briefcase down on the sofa. "God. I gotta get out of here." she repeated.

Despite the sulphur smell of the water she took a hot shower and made herself some coffee with the bottled water she bought from a local vendor. As she settled down on her sofa and turned on the view screen the thumping sound of someone in the hall, made her switch to the security scanner. There was George, George himself cleaning the puke and occasionally using a scanner against her door, trying to peek in by bouncing her video signal back to his scanner.

She knew he had videos of her naked, made by using this technique. In fact she knew every woman that lived in the apartment was part of his perverted little collection. She even tried to sue him for invasion of privacy. He got fined and ordered to destroy the tapes or give them to the persons they were of. So, he made copies and destroyed the masters. Now, Marsha wasn't a prude, or a virgin. In fact she was known as a wild child in her younger days. She knew she had a nice bod. But she didn't like the idea of this...this porker, waxing his willy to her picture. Finally she came to understand that she wasn't the only one in that situation, not that that necessarily made it any better, and she hoped one day he'd pull his pud a bit too hard and do himself a grave injury. She giggled a little at the thought of that, but the visual of George naked with that huge fat belly, was enough to put her off sex for years. Switching the viewer back to standard she watched a bit of news, a few seconds of the weather and fell asleep with the canned laughter of a sit-com running through her head.

Computer End Recording...

But I don't wanna....

Marsha had just arrived at the office after an awful morning of, well first waking up. Then a cold water shower and to top it off she spilled a hot cup of coffee down the front of her blouse. She always kept a spare blouse at work, so she removed her blouse to change, and in walked Karen.

"Well...we're dressed a bit informally this morning aren't we?" She piped.

"Oh shit...you scared the hell out of me." Marsha replied. "I spilled my coffee. I don't have a spare bra, so now I'm gonna be sticky all day long. Double cream, double sugar." she muttered.

"Oh quit bitchin...Listen, there's something we've got to talk about. When you're done come over to my office ten minutes be ok? Good...byeee." Karen said leaving almost as quickly as she had arrived.

Marsha put her jacket on over her bra, took her dry blouse and went to the ladies room. Soaking a few paper towels in hot water, she rung them out, and went into one of the stalls and cleaned up in preparation for her meeting with Karen God how she hated meetings with Karen.

As she entered the office she waited a moment for Karen to get off the phone.

"Marsh..." Karen smiled as she hung up the phone. "I just got the ok, from Corporate HQ."

"Ok? For what?"

"We're gonna send someone to the Mars II training facility on Triton."

"Why?" Marsha asked hesitantly.

"There's something going on, on Mars II that the top brass

can't understand. Unfortunately, only warriors are allowed on Mars II. So someone will be sent to Triton, spend a couple of weeks in their training program and hitch a ride. HQ had to pull a few strings and twist a few arms to get a rules change, so that a six person team can go, instead of the usual five."

"Who? Who did they have in mind?" Marsha asked half knowing what the answer would be.

"I recommended you."

"Me! You recommended me! Why the hell would you do a stupid thing like that?"

"Look we need someone there who knows procedure. Who is adaptable, someone with brains for God's sake, not a mindless warrior who can only stick to the program."

"But...but yesterday you told me the Warrior program couldn't be involved."

"Seems that's been changed. I just got the word today that some of our overseer ships stationed to monitor battles and act as referee were shot down in the pasts few months. The last one was seven kilometers outside the battle parameter, four kilometers from where any hostile actions were to take place. Not only that, it's computer downloaded sighting and lock-on information consistent with a ground to air missile, and we both know those aren't allowed. Another thing, we're scheduled to take on the Coonwaadii in a few weeks. You've read the reports, you know their reputations. If we can't keep a ref ship flying, we don't have a chance in hell."

"Let's suppose, just for a moment, that I did go. How is my dying gonna help things. Because we both know I'm not even close to being a Warrior." Marsha asked.

"You'll be assigned communications duty, base camp only. You'll act as a base operations communications link with the six ref ships in monitor stations around the battle parameter and as a backup Tech officer should they need one, I don't think they will though.

Besides, you've had field training this isn't much beyond what you already know."

"Do...do I have a choice in this?"

"Not really...I know this is sudden, and not at all what you want. But, if I can be candid you are perfect for this. You got no ties here, no family and your psyche profile shows you haven't been working well since...well since...since the attack."

"You mean since I got raped, and beaten half to death on a lame ass nothing assignment YOU should have done yourself and the investigating committee told you so but you sent me instead calling it delegating a task. I didn't have a choice then either."

"Hey...that's not fair. I gave you a choice...either go on the assignment or resign. Besides how in the hell was I to know the contacts would be members of a rape gang? Besides I though you could take care of yourself."

"Normally yes, but in seconds they had me loaded up with Tetrazine a pleasant little drug have you tried it?" Marsha asked sarcastically.

"No...I...No." Karen replied allowing Marsha to continue.

"I wish it had been a simple knock out drug, but no they wanted me awake so I could help, and without resistance I was forced to help them do everything they wanted. They dragged me into an abandoned warehouse, and began to auction off my clothes then they forced me to cut off what they "won". More like tear off when they got more excited, then when I was naked, they began to bid for sexual favors. Large hands mauling me squeezing and sucking on me so hard I thought I would die. They told me to smile, relax and enjoy, but it's hard to enjoy pain. "Suck harder bitch! Swallow it...swallow it all. Take it...take it all." I'm choking and gagging and puking and they're pressing more and more deep and deeper. "Nice tits cunt. Fuck me back...harder....harder." You've never been forced to give head over and over, swallow that crap? No, wait they wouldn't have to force you would they. Hell that would be fulfilling one of your favorite fantasies.

Well maybe it's yours but it certainly wasn't mine." Marsha yelled.

"Wait a minu..." Karen tried to say.

"Shut the fuck up!" Marsha continued; "They fucked more times than I can count. I passed out and they were still going, hitting me, choking me with their hands, their tongues and with their cocks. When I couldn't respond anymore and they had spent themselves they began to beat me, called me a worthless whore, a cunt and a slut. Then they all laughed as they left I know they didn't care if they'd killed me. I don't know how long I was out, when I woke up I was naked, bleeding from every orifice the taste of semen in my mouth, the smell of it in my nose together with blood and who knows what. I had two broken legs, a broken arm, some cracked ribs and a minor concussion. With all that I managed to drag my naked ass into the street, where a cop found me, I was drugged and in a daze but I could swear her fondled me for a minute or two and took a turn with me, before he called for backup. There's real compassion for you huh? I told the police who did it, where it was done and when, but their response was "I don't think there's much we can do about it now. If the monitors didn't see it, it couldn't have been that bad." After I recovered, those same guys were found dead and dismembered in sexually compromising and embarrassing ways. Wonder how that happened, still if the monitors didn't see it, it couldn't have been that bad.

And you with your do it or resign shit? Real comforting stuff and now you want me to do you a favor...for a woman you've got some balls you know that. Look Karen we've known each other since the third grade. I never liked you, and I can't help it if I decided not to sleep my way to the top like you did. I've done research, I've done collection and I have done deep cover, I bust my ass at this job and get that attitude from you. No...no way, if I do this, it'll cost you...it'll cost you...Big."

"What's that mean?"

"That means I want 5 Mil credits deposited in my personal account by noon today and another 125,000 credits to cover my travel expenses with an additional 5 mil credits on my return."

"I can't..."

"Can't what? Can't authorize those expenditures? Well if you can't then I suggest you find someone who can or get another operative. You have one hour to give me an answer, otherwise I walk. If this mission is as important as you and your superiors seem to think it is, and if I am so indispensable you'll meet my terms. One hour Karen one hour and fuck you very much." Marsha yelled as she left the office slamming the door behind her.

Marsha first went to the cafeteria, got some coffee and a danish, then back to her office where she plopped down in her chair and could only sit there shaking and listen to the pounding of her heart for the next few minutes. She wasn't sure exactly what had just happened, but she felt good about it. It had been the first time since the attack that she had actually gone into details. Surprisingly, it felt better than she thought it would to vent all that pent up anger and fear.

Then the reality of the ultimatum she had just dropped began to set in, and she knew a major change was going to happen either she would resign or go on this stupid assignment. If she lived through it she would be set for life, but for now the waiting game was on. She sat at her desk trying to concentrate on a proposal for a moment or two only to realize how useless that was. So she drank her coffee, took a bite or two out of the raspberry danish (eeick she hated raspberries) wrestled with her sticky bra for a minute or two, then drummed her fingers on the desk and stared at the telephone.

Marsha had made up her mind that if she hadn't heard anything by 11:30 she'd just go to lunch and not come back. It was 11:30 and no word had come yet.

"Five more minutes...I give you five more minutes Karen." she whispered still staring at the phone. Hoping she was doing the right thing. But, then the right thing is a hard thing to grasp sometimes, especially in a heavy stress situation. Funny, it seemed cut and dried to her before, but now, those retro-thoughts kicked in and she wasn't as sure as she had been before.

At 11:34 the phone rang.

"The contract is on my desk. The money is being transferred now. You have first class accommodations on the Delta Queen, dock 78 at 1:45. Everything is spelled out in an information brief you will be given by the attendant when you check into your suite onboard. Oh..ah..Marsh...that power play you did scared the shit out of the guys upstairs. They made up there minds in less than six minutes. That's a record for them...the rest of the time was paperwork and red tape. Anyway...ah...when you get back we'll throw back a few Tep's and laugh...alright. You...you take care kiddo...bye."

"I'll be a son-of-a-bitch." Marsha said as she hung up the phone. "I can't believe they fell for it...10 million credits...Holy shit!" she continued caught up in the moment. "Uh…oh, wait a minute that means I have to go...O God. What the hell am I so happy about? Then again, maybe a trip off-world is what I need now. Yeah…some new sights to see, this might not be so bad. Hell, it might even be fun." she muttered to herself not really believing a single syllable.

When Marsha got to Karen's office there was a note on the door that read: Had a meeting---The contract is on my desk. Read it and sign where indicated, encoded card is in the envelope next to the contract. Good luck K.

Marsha first checked with her bank, 5,000,234.84 alright. She then read the contract, a simple document that essentially said the same thing she had told Karen. With no hidden provisions apart from the part that said I will follow the assignment laid out in the brief. It didn't seem too complicated so she signed it, scanned the card, 125,000, yep, and left

Computer End Recording...

Assignment Unknown...

"What in the hell are you lookin' at.—That skinny, young, thing! Is that what you want a skinny redheaded child!" Were the only words Marsha could make out as she walked by the short balding man who had been caught by his wife ogling her. True the aqua-blue dress was a little tight and maybe showed a bit too much cleavage and thigh, and true she did weigh only 105 pounds. But, she never thought of herself as skinny. The child part was what got her the most though, she had turned the big Four-O her last birthday. Still, it was kind of nice to be noticed. Though any guy would have to be blind not to.

Truthfully though, she hadn't worn the dress to impress anyone, or to attract attention. Well, maybe she did want to attract a little attention. But mostly she just liked the color, and the way it showed off her bright red hair. Besides, it was her bod, and she had lost 20 pounds to get into the damn thing in the first place. Anyway, most of the time on these business cruises all a girl could find were horny business men, single guys running from something, or some asshole who wasn't ready for something serious. Not that she was looking for something serious mind you. She hadn't had anything serious since Clyde her ex took a walk after she got attacked by that gang a couple of years ago. He just couldn't take it, she couldn't have kids anymore and he couldn't take it what a bastard. Still, if the circumstance and opportunity presented themselves.

When she finally arrived at her "suite" it was really only a large room made out of two smaller ones by opening the door between them. There was a john, and a shower, both extremely small. With enough closet and drawer space for about half of what she had packed.

About two seconds after she had unpacked what she thought to be essentials and settled back to rest a bit, a light tap on her door interrupted an almost calm moment.

"Who is it?" she asked.

"It's the Purser madam I have an express parcel for you." a weak female voice whispered.

"Must be my orders." she mumbles as she opened the door and took the parcel. "Thanks." she said as she ran her card through the reader and gave the Purser a 20 credit tip.

"Thank you! If you need anything else just call me. Name's Nina...Nina Capri I hope you have a good fligh....."

Marsha wasn't really listening and unintentionally closed the door in Nina's face as she opened the seal on her orders, placed the micro-disk into her headset and pulled the visor down.

These orders are sent to you on VR/vid-chips so you may review them as often as you like. However, they must be destroyed before you set foot on Triton. You are to proceed to the Triton Warrior Training facility. There you will participate in the remainder of an ongoing training. Once completed, you and your team will be rotated on the first available shuttle to Mars II. It has been arranged that your team will be engaging the Coonwaadii, on or about the third week of next month. Your mission will be to use a multi-phased sensor array to spot any unexpected anomalies. These will be reported immediately to your Supervisor over a coded satellite link. Use code Alpha-Alpha-6ZT9 for your link. Your bio monitor is enclosed in the parcel, take it now, press the green button.

Marsha flipped up her visor, took the monitor and pressed the green button. When she had finished, she flipped the visor back down and the recording continued. The Bio monitor must be in contact with your skin at all times as it does a complete sensor sweep of its surroundings, including a DNA trace scan of your skin if it is unable to scan, or finds no match it will automatically down load the last information and deactivate. This device is an optional product and may not be used if you deem unnecessary. However, if not in uses please deactivate and destroy by placing it back in the information packet. More information is to follow also in your parcel you will find a K16 capsule, identification profile, and military background. You will use your own name, since it is unlikely that anyone off-world will know you. If you are discovered, you are authorized to use ANY means possible, to make sure the necessary information, does not fall into hostile hands.

If you are caught and are unable to do the right thing, tap the bio monitor has a small fission charge attached to it and may be activated by tapping the green button three times, fifteen seconds later a magnitude seven fission explosion will occur. To abort press and hold the green button for 10 seconds (Caution: The self destruct cannot be aborted once the 1 minute mark has been reached.) If you have any further questions use the enclosed chipset to contact Supervisor 29, however, you may only do this twice during the mission for information through the satellite downlink. Good luck Destruct sequence Delta-365 now initiated, when you wish to initiate please place disks into the parcel package, and squeeze to activate.

Marsha took the vid-chip placed it in the parcel and set it aside for a while. She'd look at it again later, to see if there was something she possibly may have missed. She then took the K16, put it into its safety container and placed the Bio monitor on the bedside table. Then she remembered it had to be next to her skin at all times. Instantly she began to research all the placed she may be able to put it. Finally she decided on her cleavage, predictable maybe, but it's for certain that no one would find it there, unless the right man should come along. In that case both of them may die, because there's no way she'd attempt to boff someone with that thing attached to her.

They hit hyper-net, just when Marsha was sitting down to her first meal. Her stomach fluttered for a few seconds, until the dampers started their series, and finally smoothed out the ride. She ordered the Jabrean lobe fish, it looked and tasted it had been dead a week or two, then marinated in forty weight, and brushed against a fire to warm the corpse. It looked and tasted disgusting, so she ate around it. The potatoes were great, and the veggies were fresh.

She had just left the restaurant when she noticed a sweet shop with Pholarian white chocolate on display. Pholarian white chocolates were probably her only real weakness, so she bought five pounds. Thinking maybe she'd come back later for ten more. For the most part her trip was uneventful, except for the time the fat Caldorian tried to pick her up. She kind of hoped to see the very attractive Kaldite man, who had been on the station a couple of weeks ago when she was on vacation, coming back from Altare's ninth planet, Soborand.

One evening as she sat at her table after dinner he sent the most beautiful lavender rosebud to her table. It was the first time she had blushed in years. But, no such luck this time only the fat Caldorian, and a few leers from other perverts. So, she spent a great deal of time in her cabin, eating white chocolates, re-reviewing the vid-chip and generally being bored.

On the morning of the third day, the ship docked at pressure dome 9, of the Warrior Training facility. After destroying the vid-chip and opting to keep the Bio monitor she disembarked, walked out down the pressure tube, where she was met by a flunky and escorted to Commander Hardesty's office. There she waited almost two hours, with the flunky staring at her, every time she looked away. Finally the commander arrived, looked at her sitting there dressed in her tight black dress. Mumbled a bit and said

"Come into my office, eh...Coronal ah...Tate."

Marsha felt a bit self-conscious so she attempted to pull her short dress down a bit. It didn't work, in fact the only thing it did do, was give the flunky a better look at her cleavage, and created for him an embarrassing problem when he stood at attention.

Marsha went into Commander Hardesty's office, where he closed the door and motioned her to a chair.

"I understand that this is your first day here...Coronal. But, I am still you're the base commander and, a salute is usually what accompanies all incoming warriors, when presenting their papers and formal introduction." He commanded.

"Beg pardon sir!" She replied as she came to her feet, snapped a salute and continued. "Coronal Marsha Tate, reporting on special assignment as ordered and my papers sir."

"Well...Coronal. You realize you have a lot of make-up work to do? You only have three weeks of training in this session unless you want to roll over into the next." He said looking at her paperwork.

"Would that cut into my rotation to Mars II?" She asked.

"Yes...It would." He replied.

"Then I respectfully decline. Circumstances demand that I be on the next Mars II rotation. SIR." She said.

"Why?" He asked gruffly.

"That information is on a need to know basis sir but if the commander will make a request of the Warrior Program Council. I am sure all of his questions will be answered." She replied.

"Warrior Council...there's another worthless organization. All they do is double-talk, bullshit, and sit around with their thumbs up their asses. Now get this Coronal, I don't like you, I don't like the Council, and I particularly hate not knowing what the hell is going on. So, here's the pitch, you have any problems, you work them out. Report to quartermaster dome 3, and pick up your uniforms. Then shag it over to shack 245 and report to Kajeem Campbell, he'll proceed from there." He commanded.

"Yes Sir!" She replied picking up her bags.

"Leave the bags here, there's nothing in them that is any use here. Unless you're one of those who think they got tits like no one has ever seen before, you can get what you need for modesty. But it'll be stored until your training is finished." He barked.

"No need sir." she said dropping the bags and going out the door.

It took her a few minutes to find the quartermaster, and she was more than a little surprised at the training...eh uniform. With supplies in hand she ran over to shack 245. No one was there so she sat on a cot and waited. A few minutes later Campbell walked in.

"What the hell are you doing sitting around when there's work to do." He yelled.

"Coronal Marsha Tate reporting as ordered." She replied.

"Ain't no Coronals, Majors or Generals in this place. You are a

Warrior maggot. You got three minutes to get into your training uniform, and out on the blue line! Ya hear!" He commanded.

"Yes Sir...But."

"But...but what? Oh dear lord do I have another baby whose shy about the training uniform. I'll get you for this Hardesty. Jees, he told me I'd be getting a new recruit. I was kind of hopin' for a guy. Then I don't have to put up with this female garbage. Look I'll explain it to you only once. This is not like boot camp on Earth, in fact it's not like boot camp anywhere else in the galaxy. We train Warriors here, not men, not women, Warriors. The reason for the construction of the uniform, is to allow maximum respiration without the restriction of clothing. So, if you're another one of those who thinks they got golden tits, that no one has ever seen the like of, then wear a bra. Your time is running out maggot."

"I don't have a bra." She replied.

"Too bad...put on the uniform and get out on the blue line. You got one minute and forty-five seconds. Now shag it maggot."

Had Marsha known that this would be a clothing optional day she would have been more conservative in her clothing choices this morning, she would have worn a bra or taken Hardesty up on his offer to take one out of her bag. But, she was told how hot it was on Triton not an exaggeration, it was HOT and the pressure was like an elephant on her chest. So, she opted for the physical comfort of going without. Not knowing at the time that the real look was naked or nearly so. She deciphered the training uniform, stripped down and put it on. When the uniform was on she looked at herself in the mirror. This would give that flunky a woody he'd have to work on for years. She thought to herself. It took another minute for her to figure out where she could put the bio monitor with most of her flesh exposed all she could do was stick it down into her training pants on the right side of her butt cheek. It didn't show; as much as the rest of her did anyhow.

"Time's up maggot, naked or not get your ass out to the blue line right now!" Campbell commanded.

As Marsha came out the door the others were standing at attention on the blue line. She took her place at the end of the line and stood out like a white sock on a line of brown. Her white skin appeared more than just a little white, she practically glowed. In fact the only thing about her that wasn't glowing white, was the deep red blush she had, when she realized that Jarob was the man, who had sent her the rosebud. And there she was standing half naked in front of him.

"Tate, name, rank and point of origin!" Campbell bellowed.

"Tate, Marsha. Coronal. Earth Warrior Program." She yelled.

"As you may or may not know, the Warrior Program has changed the format of battle on Mars II. The teams are now six instead of five. Tate, will be our communications and back-up tech officer. Welcome to the team. First and foremost, I will stress as I have so often in the past. As of right this minute, you maggots are no longer individuals, you are a team. Anyone who does not consider the team, in everything they do from now on, endangers us all. That's the speech, so are we all ready for our ten-k?" Campbell yelled.

"Yes Kajeem!" they all replied.

"Then move it out maggots." He bellowed.

The embarrassment of the moment was gone almost instantly, when Marsha trailed far behind the others, trying to run the ten-k up the rocky slopes of the newly installed track. Even Dafora felt better having this straggler around, now she didn't look so much like the weak link anymore. Marsha was trying hard, but toward the end her legs became rubbery, and she was tired of having her boobs bang uncontrolled on her chest. Holding them didn't last long, and her chest felt like every ounce of air in them, was about to explode out in one big burst. As she finished the ten-k, she fell to the ground and felt as if even the training harness was too much to be wearing. Pageena came over to her as she lay there.

"It gets easier, just give it time. Probably the biggest thing to get use to is the pressure, and temperature here. It's about what Mars II normal would be. Take long slow breaths."

In a minute or two Marsha felt strong enough to stand.

"Thanks." She said.

"Ok maggots...you got half hour free time while I go talk to Hardesty. If I'm not back by 08:30 you guys report to the ripper. Tate you take level one, the rest of you pukes take level five." Campbell barked.

They all walked back to the shack and instantly went for their cots.

"You can take that one on the end by me." Pageena said pointing to an empty cot.

Marsha lay down and tried to gather herself together, she couldn't believe how much the ten-k had taken out of her; having run a ten-k on Earth almost every day, and never felt this bad.

When she opened her eyes she saw, Jarob, Benni Jack, Dafora, and Pageena standing around her.

"Hi...I'm Benni Jack Taylor, this big dumb lookin guy is Jarob Ka."

"I'm Pageena Timbo, and this is Dafora Meps."

"Nice to meet you...where did you get the bra." she asked Dafora.

"Kajeem Campbell gave it to me until the new uniforms arrive." She answered as she walked to her cot and lay down.

"New uniforms?" Marsha whispered looking at Pageena.

"Campbell just gave her that to stop her from bitchin' all the time. She almost died the first day wearing a silly thing called an Ootman, which is customary on their planet." Pageena whispered.

"She...she's an ODII? There's never been a female ODII in the program. Someone must have fucked up somewhere." said Marsha.

"You got that right." said Benni.

"I wish I had the mother who screwed up on that one here right now." said Jarob.

"The harnesses aren't so bad...once you get used to them." Pageena added.

"Sure, that's fine with you. You're Amnan, the only time the people on your world wear clothes is when you go off-world, you uncivilized, pagans." Dafora chimed in.

"You're Amnan. I've heard about your species. If you don't mind can I ask you is it true that you...?" Marsha asked.

"Yes...we have been known to engage in what you call sex for days at a time." Pageena answered not waiting for the rest of the question.

"Days!?"

"Yes days."

"But, how can you do that?"

"Our communication is based on what you would call sex and sexuality. We link not only physically, but telepathically, whenever contact is made. We can do it with anyone, not just our own species. But most of the time we don't. Other species just don't seem to have the stamina. Jarob here wanted to screw me when we first met, as I'm sure he does you, right now. But, I would have killed him I'm sure. And to be honest any man who flaunts his sexuality so openly, must have little to offer, or the need to flaunt it wouldn't be so strong. Right?" said Pageena with a wink.

Marsha saw Jarob instantly blush and walk away. Benni Jack trailed close behind him.

"See how intimidating it can be to be not only sexually superior, but half naked and untouchable." Pageena said.

Jarob was just about to turn around and tell Pageena to go fuck herself. When Benni Jack told him it was useless to talk to her.

"Ain't it amazing how a little bare tit can do that." Pageena yelled, trying to goad Jarob.

"I agree, most men think women are a pair of tits, a twat, and no brains." said Marsha slipping Pageena a wry smile.

"Hey, that's not fair. You don't even know me." Jarob said.

"Yeah but it's been my experience…" Marsh tried to explain.

"I'm sorry if you've had bad experiences with men, and likewise I'm sure you've ran into a bunch of assholes in your life as we all have. But, do me the courtesy of at least getting to know me before you judge…ok. But…I am an asshole. Ha ha ha." Jarob laughed.

"You have a reputation on station 16. Is it wrong?" Marsha replied.

"What? What did they say about me?" Jarob asked.

"Just how sexy you were and what a great lay. You know girl talk." Marsha teased.

"Really? They…they really said I was a great lay?" Jarob asked.

"Some of them---then again some of them thought you were just an asshole, like you said." Marsha replied.

"I agree." said Dafora.

"Me too. You are an asshole." said Pageena with a broad smile. "But not TOO big an asshole."

"We'd better get going or Campbell will show us who the real asshole is, by tearing us a new one." Jarob smiled.

"Marsha, you'd better put your battle top on before we go."
Pageena instructed.

"Why?" Marsha asked.

"Trust me, don't ask." Pageena replied.

Computer End Recording...

Three more weeks...

Campbell and Hardesty had been good friends for eighteen years. They met back when two teams of four, were the normal complement. Before the Kazanti shields, and EM barriers, when running battles went on for hundreds of kilometers. Hardesty took a plasma round that should have taken Campbell. He returned the favor by grabbing and throwing a phosphor mine a few seconds before it blew. It took every bit of hair off his chest, and blinded him for almost a year. Weapons like that are illegal now, now it's just down in the dirt, close order combat. There had been changes in the team size before. But, this time was different. This time no one had been given any notice at all. Not even the Commanding officers of Triton. Officially, it could be said that no real change had been made. Since, this was the only exception. All other wars on Mars II would proceed with the normal contingent of five-person teams.

"Look Hardesty, I need to know what the hell is going on here. This sudden change in the team strength, is screwing up everything. Why six all the sudden? Why only on this sortie? And why the hell did they send this Miss Universe, warrior wannabe here, I can't have a team member who'll cry when she breaks a nail." Campbell bellowed chugging down a scotch Hardesty had handed him.

"Now...just unruffel your feathers Luke." said Hardesty as he poured them both another scotch. "You and I both know that the brass, couldn't find their butt with both hands. I've requested more information from the Warrior Program Council on sat-link, we should know something soon. But, I tell you. If this Tate babe fucks like she looks, you could have a good time with her when this is all over."

"You know I don't go in for that shit. Quickest way to make an enemy of a women is to fuck'em least without thinkin' 'bout it hard, before, during, and after. B'sides Tate seems to be a good recruit. A bit rough around the edges maybe, but she's got what it takes, I can see that. That crack about the nails was just venting." Campbell explained.

"Oh...shes got what it takes alright..." Hardesty said with a leer in his eye.

"Dirty ol' men like you don't stand a chance anyway." Campbell replied.

"Yeah...but that don't stop us from trying."

"You horny old bastard. Thanks for the hootch, and make sure and let me know when that info comes in from the Council." Campbell said as he opened the door to leave.

"Will do...my friend Luke...watch your ass, and watch hers for me, too. I mean that seriously." Hardesty said.

"See ya later you old reprobate." Campbell said with a big smile, as he closed the door behind him.

Marsha had tried three time, in the past ten minutes, to beat the five character team, in holographic simulation of level one. Each time an enemy would swing in from either her right, or left side, suspended from a line on the overhead cross beam. Then the alarms would go of, and the force field would come down, dead again.

"Reboot program L1." She gasped.

"Cancel reboot." Campbell interrupt.

"Why? Why did you do that?" Marsha wheezed.

"I've decided, that this is and unfair way to train you. How can you be part of the team, when they're on level five, and you're on one. For that reason, I have set the programmers to work, with design instructions for a six person team. The program should be in operation tomorrow. Meanwhile, you hustle your butt over to the Tech-Center. As back-up tech you'll have to understand everything Dafora does. When you get back you'll coordinate with her, to get your studies in line with where they outta be. First go back to the shack and stow that battle top or you'll die from the heat. Shag it out maggot." He bellowed.

Marsha went back to the shack and put away her battle top. She was both happy and nervous about putting away her top. Happy to be rid of such a heavy and restricting thing, and nervous to be again walking around almost bare assed. But, she had to admit it was easier to breathe without a top. She also knew that if she had worn any of the clothes she had brought with her, she would have to soak them in disinfectant for a year or two to get the smell out.

When she reported to the Tech-Center, Marvin almost had a seizure. He walked up to her and said his favorite line.

"I ain't had a woman in eight years."

"It's likely that another eight years will pass before you can remedy that situation." Marsha replied.

"Ummm...can we help you with something?" He asked looking directly at her breasts as if hypnotized.

"Ar...are you in charge here." she asked.

"Marvin, go take your pill and lie down before you hurt something." Dr. Keefru interrupted.

"I'm sorry Miss...Marvin's not always as tactful as we would like him to be. But, to his credit he is one of the finest micro-circuitry analysts in the business." He said.

"I think he's kind of a dork." Marsha said.

"Well, perhaps he is...But excuse me, I am Dr. Davis Keefru, in charge of Research and Development, of most of the Warrior technical equipment. You must be the new warrior Kajeem Campbell told me about...Coronal Marsha Tate?"

"That's right." said Marsha, turning with her back to Marvin as he continued to ogle her.

"You must excuse Marvin, I'm afraid he's a bit obsessed with breasts. Every time a female recruit comes in...Especially in training...eh...clothes. He tends to, shall we say, digress a bit." Dr. Keefru replied.

"A bit! The man's a moron with a tit fetish." Marsha quipped.

"That may be a little strong, but, I guess for want of a better definition...." Dr. Keefru replied.

"Just keep him away from me or I'll write a new definition for blue balls." Marsha said.

"Oh...never mind him...Are you able to field strip a multi-phased sensor array?" Dr. Keefru asked.

"The MS 140 or the MPS 465? And is that with or without the voice recognition circuitry?" She asked.

"We mostly work with the MS 140's, with voice recon. But we have used the MPS 465's in some situations." He replied.

"Why? The MPS 465's so much easier to calibrate and it's only a three step breakdown if you have to repair something."

"Ahh...but the 140's are better shielded from sub-space field distortions." He answered.

"Just wrap a sheet of valdite around the pole between the primary and secondary coupling. The valdite will act as a sub-space resistance phaser, and shunt all frequencies that are high or low enough to distort the scan, into the phase discriminator and recycle the frequency into the buffers along with the sensor scan reactivator. Then every time the sensor sends out its secondary signal for follow up, the aberrant signals can be added as a carrier to extend the range. If you get too much distortion feedback just do a wide field delta scan, and if that doesn't work just smack it on the back panel like this." said Marsha with a bit of a smirk as she slapped the back of the sensor box.

"That would work, but only if the feedback was naturally occurring and not an enemy attempt at disinformation." He replied.

"Even if it were the enemy, once you drop into a wide field delta, you'd scatter their scanning frequency with so many ionized particles they'd be days trying to re-calibrate. Hopefully, by then it would all be over. What code series are you using the 915's or 1165?" She asked.

"Most of the time we find the 915 series adequate, with the 465 even though the 1165 was designed for that system."

"It also takes less time to memorize. But with 475 null codes that cuts the list considerably anyway. What's the active long and short on these?"

"Active long has been augmented to 270 kilometers, short is about 92 kilometers. But, you can vary the band widths to get a better interpretive reading. You...you know the multi-phased sensor pretty well...how did you come by that knowledge?" He asked.

"I was one of several engineers who designed that particular scanner. Both of them actually, but we thought the 140 would be used as a multi-phased sensor to do planetary mapping, not warrior work." Marsha answered.

"You're...you're THAT Marsha Tate, inventor of the multi-phased sensor diode." Dr. Keefru asked.

"That's me. But I did that back when I was fresh out of College. The warrior program had just recruited me. I thought it was great at the time. But, look at me now." said Marsha looking down at her bare breasts and skimpy training uniform.

"I...I gather you know the 915 code series well?" Dr. Keefru asked.

"Know it, I wrote it and the 1165 as well. But, I have to admit I thought the Council was going a little overboard with the 1165. The chances of cracking any code series in the 915 is 145,764,286 to .1 and of course even higher in the 1165 series. But, they wanted real security, like 146 million to one wasn't secure enough." She answered.

"Then I gather this briefing would be wasted on you. But, look maybe you can help." Dr. Keefru asked.

"Sure...what?" Marsha inquired.

"We...we have this Warrior...Dafora..." He said.

"Dafora Meps. Right?" She cut in.

"Yes...she's...she trying very hard, but it's been over a week now and she hasn't made much progress. Do you suppose you might be able to help?" He asked.

"I'll try...but you know OD's. The last time they asked for help, the entire Planetary Council on Economics almost went bankrupt. Two days later, and nine planets would still be in financial ruin, including Earth. Stubborn, pig-headed bunch of pseudo-intellectuals. But, like I said I'll try." Marsha replied.

"You'll be wanting the MPS 465 then? I guess the code book won't be necessary?" He asked.

"Yes to one, and no to two. Thanks Doc...and ah...keep Marvin on a leash or he'll hurt himself. See ya later." said Marsha as she took the MPS 465 and walked out of the office.

On her way back to the shack, she met up with Campbell, who ordered her to get over to dome five and join the others for hand to hand combat. When she arrived...

"Mar? Missy Mar?" Do-Gen shrieked.

"Do...Do-Gen...Is...is that really you...wha...what the hell are you doing here?" Marsha asked as she hugged and received a large hug from Do-Gen. "I could ask you the same question." she continued.

"I've been teaching hand to hand here for about a year and a half now. But, tell me how are you? Are you still having trouble with what happened?" Do-Gen asked in a whisper with her arm on Marsha's shoulder.

"Not so much any more. But, shit what in hell prompted you to come to Triton. You could have had your pick of assignments?" Marsha asked.

"Well...I thought a little off-world time might do me good, and the Warrior Program is always looking for people willing to tough it out, way out here at hell's asshole." She replied.

"So tell me, did you bring Zuo-Sho or is that mountain back where the real people are?" Marsha inquired.

Do-Gen was just about to answer, when Zuo-Sho crept up behind Marsha, and grabbed her roughly by the neck. Marsha pivoted on her right foot, spun left and caught Zuo square in the chest with a full contact spinning reverse kick. Zuo let out a groan as he flew involuntarily across the room.

"Oh...ooh my God! Zuo...Zuo are you alright." said Marsha as she quickly ran to Zuo's side.

Zuo sat up, shook his head, and looked at Marsha.

"You're still telescoping your kick, keep that up and your power will be useless, your contact control could use some work too." He said with a large smile beaming across his face.

"You should know better than to sneak up on me!" Marsha replied.

"How else can I keep you on your toes?" He asked.

"Surely there's a better way than getting your ass kicked." She replied.

"You know it would take a lot more than that skinny little leg of yours to hurt me....Hey...hey Marsha you...you're filled out real nice...funny I always thought you were flat chested." He said with a huge grin on his face.

Marsha had forgotten she was wearing her training harness, and for a moment was self-conscious. Then she looked around and noticed that everyone was wearing the same thing.

"But, my boobs are prettier than yours." she smiled looking into Zuo's loving eyes.

"I agree with that." He replied as they hugged.

Marsha had met Zou and Do-Gen on a mission. One of those

covert op things that usually meant someone higher up was insecure about something. Zou saved her life more than once and Do-Gen nursed her back to health and started teaching her Zen-Qui. Both were like second parents and it was apparent that love still existed between them. She lost contact about two years ago and had been more than a little hurt at not having them there when she needed them.

"How...how did you hear about the...the."

"The attack?" Do asked.

"Yes."

"Before we left to come here we asked to see you. Your boss told us you were in the hospital and that you had been attacked by a gang and hurt very badly. We tried several times to contact you but you wouldn't answer your phone. We left a note with your boss Karen, and thought you would get back to us in time. When you didn't we...we couldn't wait any longer and it was either lose our commission here or take a shuttle. I'm sorry...we needed the credits." Do explained.

"You know if we had known you needed us we would always be..." Zou said.

"Dear...dear Zou...and Do I know...I know...I love you both very much too." Marsha replied calmly hugging them both again.

"Ahum...I hate to break up old home week...But, could we get started before Campbell comes back and makes us run another ten-k for loitering." Jarob interrupt.

"He's right." Marsha admitted; "Let's get to it."

The lesson resumed, and Jarob found that he had a new respect for this new recruit. She could take care of herself, was bright, if a bit over confident, and he was sure she would be a fine addition to the team. Benni Jack and Dafora were not at all impressed, when they first met her. That opinion changed however, when they saw how easily she managed a takedown on Zou. Pageena, for some strange reason, seemed to take to her right away.

Mars II

Computer End Recording...

A World of hurt?...

The new program for the ripper had been installed and the team had just finished its morning ten-K. Although Marsha hadn't been in the field in almost two years, much of the training she had when she first became an agent at the Office of Intergalactic Intelligence was similar, it was good she remembered it, and nice to know she still had the physical endurance to keep up. She knew her role in this team wasn't one of actual fighting. But, if the time should come for that, she felt good knowing she was ready.

With less than three weeks of training to go through Campbell was still a bit unsure of the reliability and competence of his team. But he had seen marked improvements in everyone. Dafora and Marsha worked unceasingly on the codes for the multi-phased scanner even though Dafora had missed by almost a week and a half, the time she was given to commit the code book to memory.

When they entered the ripper, all were amazed at the complexity and sophistication of the program. It was detailed right down to the hairs on the inside of the monkey's ears. The battle zone now lay before them. They each took three steps before the attack. The first to destroy a target was Jarob. In subsequent seconds the entire opposing team was terminated. After nine other scenarios were completed Campbell checked the system to make sure it was operating according to specs. There was something wrong, this team wasn't that good yesterday. But nothing was wrong with the equipment. Campbell smiled; something the team had only seen him do once, twice maybe. After the team won another six stages, he practically beamed, even though he tried very hard to hold it back. Pumped up and wanting to continue, the team was about ready to start level 16, when Campbell stopped them.

"You'll need your special weapons beyond that level." He said. "We'll be doing something different from now on. After chow, I want you all to report to zone twelve in full battle gear. That includes all of the weapons and armor in your kit. For the next three weeks you will hump that equipment with you wherever you go. Now you tech's and tach will have the option of removing one piece of equipment from your pack so you'll have a place for your specialized items. It cannot be either the plasma rifle, dieapoit, or the blaster. However, the nygor, mobor, or the shield can be omitted. If you so desire and can find enough room, you may choose not to omit anything. All training gear is to be taken back to quartermaster...Dafora, I'm sure that won't break your heart. Battle gear only from now on back to the shack, shower, grab some chow and meet me at zone twelve in one hour and fifteen minutes. I got a surprise for y'all." He continued with a wry and devious smile.

Jarob was the first back in the shack, when he opened the door...someone had been in the shack...no maybe something. As he stood there staring into the room. The others joined him and they too, were taken back by what they saw. The shack...it was...different. In fact it was almost pleasant. The cots had been replaced with real honest to God beds...with big fluffy pillows...and sheets and warm fuzzy blankets. A partition now surrounded each bed giving each occupant twelve feet of privacy when the partition was sealed. The lockers had been replaced with standing cedar closets, and the showers had been individualized to ensure privacy. At the far end of the shack was a Byzalian liquid pool table, a gaming table, and several view screens. A food/drink replicator had been built into the room just to the left of the new game room.

"If there's anything you need or would like to have, just give me a request in writing and you'll have it within 3 days." Boomed Campbell's voice from behind them; "I...I had to get a look at y'alls faces when you came back here and saw this." He chuckled; "Hey...you ain't maggots no more. Warriors...you're all top of the line warriors now. Training from now on will consist of two ten-k's a day, weapons and tactics, and getting caught up on the briefings you missed while you were maggots. Looky...here, we got vid-screens and a library of movies going all the way back to 2176. The replicator is programmed with over 250 dishes and 500 beverages from each of your home worlds. Of course it is capable of being programmed for more. So, if you have a favorite dish that ain't on the list, call the programmers. Here...here's the best part." He continued pointing to another small room; "We're gonna be roomies."

"What?" They all said in unison.

"For the final three weeks of training, it is important that I get to know each of you, and likewise you get to know me. I'm a member of this team, not a Kajeem anymore, a team member. Up to now I've been in charge, from now on it's time we really work as a team. I'll bust my ass for the team, just as long as I know they'll do the same for me."

"Why the change?" Jarob asked.

"You didn't think that the maggot thing was for real or permanent did you?" Campbell asked.

"Hell I don't know. I just kind of thought it was part of the program." Jarob replied.

"Hey, we treat our warriors very good. You think this is nice. Wait till you win your first chit." Campbell answered.

"What's a chit?" Pageena asked.

"Well when you win a war on Mars II, you get a chit, It's only a ribbon really. Yellow is stage one, green stage two, blue stage three and red stage four. Anything above that is silver. When you get 3 silver chits you retire. After Seven wars, you win the big prize." He replied.

"What's that?" Jarob asked.

"Beer...lager cold..." He commanded the replicator. "...freedom a house anywhere in the galaxy and enough credits to see you through the rest of your life." Campbell replied as he took the beer and sat in a large comfortable-looking chair in his room.

"You've been to Mars II haven't you?" Benni asked.

"Oh, hells yeah...five times two more and I'm gonna go to a little planet I went to once in the Pleiades. It's got fresh sweet water and the best ground for farming in the entire universe." Campbell replied as he took a sip of beer.

"Funny I don't see you as a farmer." Pageena replied.

"Well, I guess I can see where you got that from. But I really am a simple guy. I've wanted a farm for a long time. Maybe I just wanted peace for a long time. Either way I'm almost there. My wife Jenna will be happy about that."

"Jenna? That was my mother's name." Jarob replied.

"Yeah I know. I read your dossiers remember."

"God, you're married too? I guess there is a lot we don't know about you." Pageena said.

"Got any children?" Dafora asked.

"Yeah...a son Hupak, he's seen 8 turns, and a daughter Marillia she's seen almost 17. Hupak told me last time I saw him, he wanted to be a warrior like me, when he grew up. I told him he didn't really." Campbell replied holding a holo-portrait of them.

"Jeez married, two kids, what the hell did you get into the program for?" Benni asked.

"You ever heard of Mazook station?" Campbell asked.

"No." Everyone replied, except Marsha who was quiet.

"Well?" He asked looking at Marsha.

"Mazook station, was a small colony of rebels that held out against the Dowsa for almost a month. That happened almost three centuries ago, you don't look old enough to have been there." Marsha replied casually.

"Nalvoron's rarely do." He replied.

"You're Nalvoron?" Marsha asked.

"All my life." Campbell replied lighting a cigar.

"That's impossible, the entire Nalvoron species was destroyed a week after the Mazook incident." Marsha replied.

"My family and I were buried under the rubble on Mazook station for almost three weeks. When they found us, my wife and I were alive. My mother, father and two brothers, weren't so lucky. Seems Dolirites, have this thing about Nalvoron's." Campbell said.

"Nalvoron's? Dolorites? What the hell are you two talking about?" Benni asked.

"I have no idea." Marsha replied; "The Nalvoron's went to war with the Dowsa over three hundred years ago. Because the planet was vaporized, it was assumed that there were no survivors. The Nalvoron's had no inter-planetary travel in fact it was also assumed they had no technology at all. I guess those assumptions were incorrect."

"They were indeed. All of our space ships were hardly any bigger than a large cat. The reason being that they had only to take our teleport pods to the planets we wanted to go to. Then once the pod had been delivered, we could teleport there anytime. Then quite by accident, a malfunction forced us to land one of our ships on Dowsa. The Dowsa took that to be an act of war. Then again the Dowsa took just about anything to be an act of war. Within two of your days, our planet was issued an ultimatum. We were given a certain time to surrender or be destroyed. Since we had no weapons of any consequence, we began a planet-wide evacuation, hoping the Dowsa would come to our planet, find its population gone and leave. It didn't work that way. Somehow they picked up on our transporter traces and began systematically going to each of the planets we transported to, and wiped us out."

"Setting off Galactic War I in the process. Because each planet you went to became a target." Marsha broke in.

"Yes...unfortunately that's true. We realized that, so we decided in the limited amount of time we had, to head for planets that had no indigenous populations of an order higher than a plant. But most of the planets we had set pods on were inhabited, or colonized. So, in a final attempt to go somewhere safe, and not endanger others, we established Mazook station on a small planet at the far end of the Galaxy. The Dowsa found us more quickly than we thought they would. My wife and I had just arrived and met with my family, when the first wave came in. In the scramble, my wife and I got separated from my family. We had been pushed into a small room off the main shelter, when the blast hit. I don't know how long we were out. But, when we woke up it was all over. Mazook station was a pile of burned out debris, with only the Dolorites moving around eating everything in sight. We found what was left of my family a little while later, Dolorites. Goddamn Dolorites had got them all."

"What the devil is a Dolorite." Benni asked.

"A worm, an omnivorous worm about three meters long with teeth like an earth shark. They're pack animals, frenzy feeders, hunting in groups called slooges. Damn things will eat anything, and I mean anything. I even saw one take a bite out of an ore truck. The most terrifying thing I've ever seen is Dolorites rolling in under the ground, coming up under a crowd of people. Seventy-four people of a group of one hundred and three were consumed in seconds." Marsha replied.

"Where'd you see that? They only live in a few placed. Folvu where we established Mazook, Wuu-Polatius, Jival Docerus II some on Mars II and a few other places." Campbell asked.

"Oh...I been around." Marsha replied coyly.

"Anyway, when we found out that our home planet had been destroyed. We sought sanctuary on Earth colony twenty on a planetary moon called Phobos."

"Well that explains a few things about that attack." Marsha interrupted.

"Shortly after we left we found out that the entire colony had been wiped out by the Dowsa. We still feel guilty about that." Campbell continued.

"That still doesn't explain how you could be there if it happened over three hundred years ago." Jarob said.

"Well, our life is in turns not minutes, or hours, or years. At the end of each turn we regenerate. Essentially, we have to be completely decimated to die in your terms. Every organ is totally regenerative. One turn is equal to ten of your earth years. I have seen fifty-one turns. So, in your terms I would be over five hundred year old." Campbell explained.

"Then your son is sixty, and your daughter is one hundred-seventy. Right." Pageena asked.

"That's right." Campbell answered with a laugh. "I don't know" He continued. "how many of my species survived. But I was happy to see the end of the Dowsa's tyranny in my lifetime. After the war we went to Earth, told them our story and I got assigned here. Jenna is working on station 16 and I see her as often as I can. The kids seem happy, but I know station life ain't no way to bring up a kid."

"So, how did all that qualify you for the warrior program?" Jarob asked.

"To be honest, it didn't really. I told them I wasn't a warrior. But, they thought anyone who'd been through what I had, had to have been a good fighter. Not to mention the fact that there were still a few bands of renegade Dowsa around. I think they were looking for a safe hide-out more than anything else. Then, I found out how good I was at this and decided to pursue it further. Next thing you know I'm Kajeem Campbell and that's about all I can tell you." He replied.

"So, you've been here all this time." Pageena asked.

"Here, Mars II, or Station 16 with my wife and kids. I quit the program for a while after my third son, Mokarus, died of a molecular infection. He was only 2 turns. When he died, we weren't sure what to do with ourselves. I sold used shuttles and transports for a while. You know, trying to give my family a normal life, whatever that is. Anyway, that didn't last, and I came back here, cause this is, after all, all I really know how to do."

"And you think you'd be happy being a farmer?" Jarob asked.

"Hell, I don't know. But, I know I was happy being a farmer once." He replied.

"That was a long time ago." Pageena said.

"For someone with infinite time, what does a long time mean?" Campbell asked.

They all sat quietly for a moment or two.

"What about you Pageena? Your home world has no crime to speak of. What prompted you to join the program?" Campbell asked.

"I guess I should first explain something about my world. We are not a bunch of godless heathens, as many would believe." Pageena said looking at Dafora; "True, we are a nudist planet. But the planet-wide mean temperature is 103 degrees. Most of the time clothes are not only unnecessary, but uncomfortable. Our primary means of communication is based on what earthers and other planets inhabitants call sexual contact. For lack of a more understandable word I suppose that reference will suffice. Though we do not look upon it as sex per-se I had jobruna or first touch with my father at the age of 13. I don't know what age that translates to on other worlds. It is a custom of my world that the first communication on that level be done with one's father, or if he is not there with a brother, or some other male kin. The Jobruna is part of Lo-attec-jaar, or age of understanding. For young men it is usually their mother, sister or aunt to share the moment with them. I have read in books from Earth, ODII and other worlds that what we do is considered obscene, immoral or just plain bad."

"But most other worlds cannot understand the importance or pleasure the understanding brings us. In the instant of connection, we learn all that we will ever need to know about our culture, our past, and our lives. Our minds merge not only on a singular level, but on a planet-wide scale. We know everyone, everyone knows us. As the communication, and sex, intensifies, we become caught up in the single mind. This sex as it is called has nothing to do with reproduction. Reproduction is only a matter of controlled ovulation for Amnan females anyway. Being telepaths we can upon contact with any species read thoughts, and emotions. You know a touch of the hand, a kiss and the like. But it is the sexual link that we find most useful the sharing of our bodies allows the mind to be open and free of negative emotions. We can link with another species, and instantly learn all there is to know about them. For instance, if I were to link with Jarob, I would know his total life experience and that of his father's and his father's father, his mother's and his mother's mother and so on for dozens of generations. Usually, duration of stimulation governs the amount one learns. I have no organs for sexual stimulation in or around my vagina. The pleasure derived from sex for all my species is in the link and the learning. I have had sex with earth men before, and unfortunately have found that the link is weak. A muscle in my vagina contracts around the penis and the link is established. Sometimes, caught up in the communication, we forget the frailty of other species and...Well, frankly we can cause damage. Usually not permanent damage though."

"As you might well understand, communication on this level can be taxing. Also, in time an Amnan feels they can no longer learn or grow within the confines of there planetary boundary. So, we go off-world, link with other species, return and relate, and go again. I guess you could say that's why I'm here. My planet has never experienced an emotion as strong as hatred. We have felt it in other species, but wish to explore that emotion further. War is the ultimate expression of that emotion. So, I was chosen to join the warrior program at an off-world colony and experience war first-hand. We have done this may times for things we did not understand, and wanted to learn more about." Pageena said.

"But I though you said you knew and understood something the moment the link was established." Dafora asked in wide eyed fascination.

"We know and understand the images in our minds, but without a frame of reference, we cannot have real understanding. If I were to tell you the chair in the corner is red and you did not know what a chair, corner, or the color red was, you could not associate any of your reference guides to that phrase. That would make the statement meaningless. We have the texture of hatred and war, but none of the substance. That is all we ever will have, unless we have someone from our world experience it." She replied.

"So I guess you could say you're in this just for the experience?" Jarob asked.

"I suppose." Pageena answered.

"But didn't you have to experience those emotions to get into the warrior program?" Campbell asked.

"No, not really...I understood death, and the fear associated with it. So, you might say fear is what got me this far." Pageena answered.

"Then you've never experienced hatred?" Benni asked.

"No only fear, love, jealousy, anger, and death." She replied.

"How do you experience death, without dying?" Dafora asked.

"We are telepathic without contact to some extent as well. I've read in the minds of others the last act of life. I can only assume that to be death. It's all rather vague anyway." She answered.

"I suppose because I had never really talked to an Amnan I believed them to be the things I was always told they were. Your species is far more advanced, and interesting than I ever could have imagined." Dafora replied.

"I think you will find. That knowing and understanding, requires patience, time, and interest. Perhaps, it is only now that an interest in my species sparks a fire in your mind. That allows the door, once closed so tightly, to open and allow true vision to occur. If you do that more often I don't think you will like your home-world's closed minded policies when you return." She replied.

"Perhaps, but true understanding goes both ways. Right?" Dafora replied.

"Right, and now I understand." Pageena replied with a smile.

"Well, I hate to break this up. But we have a battle practice over at the green quad in twenty minutes. You can return your gear on the way...Come on folks." Campbell said.

Mars II

Computer End Recording...buffer full...

BEGIN TRANSMISSION...

END TRANSMISSION.

The Face of war.

Schaatook Vaa, the leader of the Ypsilanti warriors, stood on the rocky cliffs of Mars II, and gazed out over the battle zone just two kilometers on the horizon. Straining his eyes, he could barely make out the Bohabak campfire smoke some distance over the small foothills to the east. Herds of Jopro where grazing on the open plains directly in front of the rocky outcrop that jutted abruptly from the cliffs on which he now stood.

There were nothing like Jopro on his home world; odd looking creatures with horned spines, a long moose-like nose, big floppy ears, and eyes that seemed to reflect all the sadness that this world had imposed. Though they had four legs, usually they hopped on only two. They possessed long thick tails like that of a Pactroll on his home world, or a kangaroo on Earth. It acted not only as a stabilizer, but as another means of gathering the long grasses on which they fed. A long thick coat of heavy fur, that hung in great clumps made the beast that stood over six feet at the shoulder appear more vicious than it actually was.

Jopro and the grace with which they managed to move in such an odd and frantic manner, were not however paramount, in his thoughts at the moment. Shaatook Vaa knew the battle that lay before him and the other warriors of his clan would be long, and hard, and would see one, two, or perhaps all his comrades dead. But the dispute over the Varashuk asteroid cluster had already taken too many lives and he would be happy to at last see the end of it. They had tried sharing the cluster, but when huge new deposits of Darburite were found, greed entered the picture and the Bohabak claimed all asteroids where the deposits were found in abundance, effectively cutting out the Ypsilanties.

"So now we go to war over rocks." he muttered softly to himself. Too old for battle now, his roll of Tech was less than glorious, but at least kept his honor in the fight. He did not agree with the final decision to go to battle. But honor and battle were his clan's creed, and when the Ruling Counsel of Elders declared battle the only way out, he had little choice but to accept the commission. Still, he had assembled the best his planet or allies could muster. Even with the short training time he was confident that victory would be his. The sun began to set behind the eastern mountains and the smoke in the distance took on a blue haze as he stood a moment or two more, then turned toward camp.

"Wa moos! Da ito dook sham dook?" A guard at the parameter of the camp yelled. Requiring a password for entry.

"Dook a ito ba." Shaatook Vaa replied.

"Mashook Shaatook Vaa huumperak implaa. Varrishinka!" The guard replied as all the camp stood at attention.

"Tune all universal translators to English now. Or else the referees won't be able to understand us and we'll be disqualified." Shaatook Vaa ordered; "In a few hours we will be in the glories of battle. Some, or maybe all of us, will die, but, remember always, death in battle is not death, it is honor, for us, and our families. Sleep now and dream of battles yet unfought, and the glories we will achieve."

"Referee 1 to Ypsilanti base---Referee 1 to Ypsilanti base, over." The com system chattered only moments after Shaatook Vaa awoke.

"Referee 1 this is Shaatook Vaa, Ypsilanti base. I read, what is it, over?"

"Ypsilanti base---be on alert that the Bohabak are not responding from their base camp and their transponders are not visible on our screens. We also do not read any sign of them within the battle zone coordinates, over."

"What do you advise, over?" Shaatook Vaa asked.

"Please remain—your location until otherwise authorized. The referees will quorum and get back to you with a decision if the Bohabak are found in violation, over."

"We will comply, over and out." Shaatook Vaa replied, with a furrow appearing on his second brow.

Shaatook Vaa knew that the war would be over without a fight if the Bohabak were found in violation. However, he also knew that the Bohabak were famous for their ambush setups and sneak attacks. But, even they wouldn't be foolish enough to attack outside the battle zone. Such an action would mean instant disqualification; neither side would risk that. The conflict had been going on for too long for it to be jeopardized or lost over a simple thing like a violation.

"This is referee 6 calling Bohabak base, we realize and understand that you may be experiencing a communications blackout, and can receive but not transmit. If that is the case, you are to fire one red flare from your neutral area. If we do not see the flare within ten minutes from---now, we will be forced to consider you in violation of the Mars II War Act of 2195, section 22 paragraph 104b. Which reads quote:

All transponders and communication lines must be open to the referees at all times. In the event of communication disruption during an event, the combatants are to retire to neutral zones until such time as communications have been restored. If communications are lost prior to the events appointed starting time, said time will be halted until communications resume…end quote.

You now have 8 minutes 47 seconds to restore communications or you will be considered in violation and the Ypsilanti declared the winners by default, over."

"Referee dispatch this is referee 3, dispatch an observation squad to zone 29, level D, on grid 40. Bohabak are not responding to communications and their transponders are no longer active. We are sending you to their last known location, request you view and report, over."

"Referee 3, this is dispatch. We are complying with your request, OS 17 is maneuvering now and should be in position within three minutes, stand by, over."

"Referee 3 standing by, over."

"Referee 6 this is referee 3, hold on violation time at 7 minutes

14 seconds, stand by for verification, over."

"Referee 3, we are holding at 7 minutes 14 seconds, over."

"Referee 6 to Ypsilanti command, come in over."

"This is Ypsilanti command, come in referee 6, over."

"We have a hold on violation time, as Observation squad 17 reconnoiters. Over."

"Referee 6 I must protest that action. The rules state clearly that the amount of time to restore communications must be complied with, that time has been given and there has been no reply. Over." Shaatook Vaa said.

"If you will review your rule book, you will see that it is the decision of the referee that determines all times, and it is our decision as to what action will be taken at what time. Stand by. Over."

"Referee 3, this is referee 6, what is your status, over."

"We are still standing by for a report from OS 17. They should be at station keeping over the zone in a couple of minutes, over."

"Referee dispatch. This is referee 6. Would you give me a conference patch with OS 17, referee 3, and Ypsilanti command. I want to make sure everyone is in on the findings, so no foul can be ruled, over."

"Referee 6, this is dispatch, you have conference, over."

"OS 17 this is referee 6, report position over."

"Referee 6 this is OS 17, we are 1 minute 09 seconds from IP and are in the cycle for station keeping, over."

"OS 17 keep your line open, over."

"Referee 6, that's a roger, line open, over."

"Ypsilanti commander are you on line? Over."

"Yes, referee 6, I am on line, over."

"Referee 3 do you copy, over?"

"Referee 3 is on line, over."

"OS 17 to conference, we have a reflection, in level D, grid 39, zone 29. Request permission from dispatch to investigate, over."

"OS 17, that's a clear, you may proceed, over."

"Dispatch...dispatch...we are under fire...repeat we are under fire...grid 39, zone 29,

level D, confirmation of an air to ground incoming. Turning 180 degrees, nose up to 45 degrees, accelerating to mach 4.....We can't shake it....repeat we can't shake it...impact...we have impact...abort all functions...abort all functions...Eject! Eject! Eject!"

"OS 17? OS 17? Can you identify attacker? OS 17? Repeat can you I.D. attacker or coordinates."

"Dispatch this is Commander Duram on remote. I managed to eject. Ensigns Taylor and Graves are gone. When the missile impacted aft, it must have damaged their eject system. I'm floating down now, and figure I'll land about twenty yards from processor 23. I have major burns on my hands and legs. I don't know what my face looks like, but my visor is shattered. What ever hit us had a pyrotechnic charge. Help...please help."

The transmission then faded into a meaningless, jumble of static, white noise, and background. Repeated attempts to re-establish contact with Commander Duram were unsuccessful.

"Dispatch? This is referee 6. Did you get a coordinate reading on OS 17 before it was attacked?"

"Last reading on the recorders is grid 39, zone 29, level D, kilometer 5. What the hell is going on out there? That's the fifth OS we've lost this month. No one on this planet is supposed to have ground to air capability. As chief of dispatch, I will no longer send out OS's unless the target can be I.D'd through scanner sweep."

"Dispatch, this is referee 6, I agree. Looks like another 212. Ypsilanti command, your war is postponed until the Bohabak team can be accounted for...Ypsilanti command? Ypsilanti command this is referee 6 do you read? Ypsilanti command come in...come in!"

Computer End Recording...

Ten days

Campbell's team had just completed their morning 10 k, the first one in full battle gear. The added weight of the packs and heavy leather armor, made the oppressive pressure and atmosphere even more difficult to deal with. Pageena almost didn't make it the full 10. But, with the assistance of Dafora and Jarob, she managed to stagger to the end. Campbell suggested she take an oxy 9, oxygen augmentation tablet before she ran again. It would help her body process oxygen faster, and decrease arterial compression caused by excessive exercise at Mars II pressure levels. For that matter oxy 9 had been recommended to all the warriors in training but most of the time the acclimation to the pressure was easy enough so augmentation wasn't necessary. Still, there were some, whose physiology would not easily adapt, hence the reason for the development of oxy 9.

"Warrior Tate, report to Hardesty's office immediately. Repeat Warrior Tate, report to Hardesty's office immediately." The intercom boomed.

Three loud knocks on Hardesty's door shook him to his boots.

"Come." He commanded.

"Warrior Tate reporting as ordered sir." Marsha said as she entered and snapped a salute.

"Tate. I got an express pack here for you, and I don't mind telling you that I hate, hate being someone's errand boy." He yelled.

"Yes sir. I understand sir." She replied.

"It's marked for your eyes only. I'll be happy to leave the room when you open it. But, get this Tate, I don't like cloak and dagger shit, and furthermore I'm not sure if I even like you. So when you finish, you will remain in my office, and you and I will straighten a few things out. Right?" He bellowed.

"Sir, yes sir." She bellowed back.

"Here's the package, you got twenty." He said as he shoved the package at her and left, slamming the door behind him.

Marsha opened the package, enclosed she found a vid-chip and a small brown rectangular box containing a code key.

"Sir?" She asked, sticking her head out of the office door and looking at Hardesty.

"You done already?"

"No sir. I was wondering do you have a vid-view visor?" She asked.

"Top shelf next to the door on the right." He replied.

"Thank you. I'll just be a minute." She replied with a smile.

When she had located the vid-viewer she placed the vid-chip inside and placed the visor on her head, pulling the heads up display down over her eyes. The information contained herein is to be considered MOST SECRET. View clearance must be red-code red-blue-blue. Beginning retinal and DNA scan now. As a soft light scanned her retinal pattern, a small electrical charge read her DNA sequence. Name: Tate, Marsha C. agent 4211-KA6Z, Office of Interplanetary Intelligence, no code name or alteration required. Cleared and confirmed, proceed with information.

In a few seconds the image of Karen appeared to stand before her.

"Marsha, there have been new developments on Mars II. Headquarters thought it best that you be notified immediately. Though you will not be able to ask any questions, should you require further information, a special contact sat-link has been approved for you to use. But you can use it only once. The code key for it accompanies this vid-chip. Listen carefully. Yesterday at 07:13 am Mars II time. The Ypsilanti and Bohabak Warriors were to have competed. At 07:17 the Bohabak would not answer the referee's queries before the battle was to begin. Thinking there might have been a communication problem the referees requested a manual response in the form of a flare, which is normal in the case of loss of communication.

There was still no response. In keeping with the rules of war, an Observation Station, OS for short, was sent to reconnoiter grid 40, zone 29, of level D. At 07:20 in grid 39 of level D, at zone 29, the OS was shot down by a ground to air missile. I am sure you are aware that no weapons of that kind are allowed on Mars II, and if the weapon was fired by the Bohabak, it is possible they are intent upon annexing Mars II. The Bohabak Government has of course denied any knowledge of the OS downing, and had repeatedly stated that they allowed no such weapon to be brought with their team.

The paperwork for your battle with the Coonwaadii has been expedited and cleared. Your team will be assigned battle duty there within ten days. We have managed to set your coordinates, relative to those used by the Ypsilanti and Bohabak. Once your base camp has been established, you are to contact your sat-link supervisor and give grid coordinates, level, and zone along with a corresponding transponder signal. You will set up a marker beacon, which will display your location at all times, using the DNA trace scanner you received in your initial orders to set up a monitoring scan. Though it is not recommended that you attempt to uncover a launch sight for the missile in question, it is possible, in a casual search of the associated area, you may be able to find a burn, or some trace of a firing point. If a point of origin is discovered, you are to report the coordinates and leave a sequence enhancer, calibrated to 0.03476 at the launch sight. If no launch area is located, do nothing.

This is the third time in as many months an OS unit has been fired upon. All have been ground-to-air assaults, and all have had the final result of both the disappearance of the two combatant teams, and the loss of one or more lives on the OS units. As an OII agent you are to observe and report anything out of the ordinary. If you find you cannot complete your mission parameters, without disclosure of your mission to your team mates. All team members are to be considered expendable. If any information about your mission is leaked, you are to use extreme prejudice in their elimination. This vid-chip is to be destroyed when the briefing has ended. Office of Intergalactic Intelligence JFR-1005977-4661 #000185, red-code-red-blue-blue. File #67 BZET Alpha 919. END UPLINK---END TRANSMISSION.

As the display went blank, Marsha could hear Hardesty

opening the door to his office.

"Well?" He said gruffly; "You done, or am I to move out and let you move in?"

"I just this moment finished." Marsha replied.

"Good...Now suppose you tell me what the hell this is all about." He commanded.

"Sorry sir. I can't do that." She replied.

"Why not?"

"You'll have to contact..."

"Yeah...yeah I know. I'll have to contact Warrior command. That's bullshit and we both know it. What if you say...left that vid chip here and well...forgot you left it...you know...You couldn't be held accountable for that could you."

"You may be right." Marsha replied handing him the visor with the vid-chip still in place.

Hardesty then placed the visor on his head and pulled down on the heads up display. Two seconds after the security display he received a hard electrical shock to his temples. Not enough to kill him, but more than enough to give him one hell of a headache.

"Your DNA sequence and retinal patterns are wrong for this security code. Just be glad that the transmission was a level 3, anything above that has a lethal gamma dose. Security vids can't be copied, or viewed without a security match that is imprinted on the chip. Sorry." Marsha said as she picked the visor up off the floor where Hardesty had thrown it, removed the vid-chip and crumbled it into small pieces.

"A security vid, that's intelligence hardware you an operative or something?" He asked.

"If I answered that, I'd have to ...you know." Marsha said with a smile as she opened the door and was about to leave.

"I haven't dismissed you yet Tate!" He commanded.

"Look Hardesty, we both know that I am involved in an operation that is not strictly military. And we also both know that if I tell you any more, then the guys in the delta blacks with big muscles are gonna come in here and burn this place to the ground. I can't tell you anything beyond the fact that there is something unusual happening on Mars II. Even that is more than you should know. So, do us both a favor and understand that if you do breathe a word, I can, will, and must kill you. It's been real...bye" (Marsha walked over and placed both hands on his desk as she spoke and stared right into his gray-blue eyes.) She then turned, smiled wryly, and walked out the door.

When she arrived back at shack 245 the word had already gone out that her team was to meet the Coonwaadii in ten days.

"We got a briefing in twenty, over at dome 6." Campbell said after he circulated the papers containing their orders.

"Ten days? That's kind of quick ain't it?" Benni asked.

"A bit maybe but it ain't all that unusual. Maybe the Ypsilanti and Bohabak battle didn't take as long as they scheduled for." Campbell answered.

"Yeah...the Bohabak ain't exactly known for their strict adherence to the rules anyway. Maybe they got disqualified. If that happened then they lost. Boy, I'll bet that made them real happy." Pageena chimed in.

"You know it. Man I can see ol' Aperia Keczo now. Eyes bugging out, all four nostrils flaring and the stiff bristly stuff they call hair, sticking straight up." Campbell laughed.

"How do you know a Bohabak warrior?" Pageena asked.

"I met him once at Station 16." Campbell said pulling his boots on; "He ain't just a warrior, he's the clan leader and probably the leader of their whole damn team. Hates losing, hates it! I saw him take on twelve Caldarians in a bar fight there. When he reared up and puffed out his chest, all eight foot three of him, with that bristled hair standing straight up, then he started bellowing. Man, I thought sure those Caldarians would piss their pants. I almost did and I wasn't even involved. But those Caldarians, well they ain't more than three foot nothin' you know. But, they kicked the livin' shit out of him and threw his ass right outta that bar. It was the funniest thing I'd seen in years. 'Course no one had the balls to laugh about it, not in front of him anyway. He came back a couple o' day later with a quanzite bomb and blew them and their ship right to hell. Yep, six minutes after they left bay 9 on their way home and BAAAHOOOOM all gone. Station security had an idea who did it, hell everyone knew who did it, but no one could prove a thing. So, they told him he'd worn out his welcome and sent him home."

"What about the Ypsilanti, maybe they won?" Dafora asked.

"Maybe… maybe Hell, there ain't no way we can know for at least a year anyway till the war reports come out right Campbell." Jarob said.

"Yeah, and I personally don't give a damn one way or the other right now. The deal is we got the duty now. Mars II, it ain't fun boys and girls. It ain't no fun at all. Be advised you better pay close attention to all the briefings from now on. You'd be surprised how many battles have been lost on account of small stuff like what you get from the briefings." He replied.

Making there way to dome 6, they entered, sat down and waited. A few minutes later a vid screen began to lower, and once again Director Coorbas began to speak:

This briefing is designed to give you an overall look at Mars II. In it you will be given more detailed history, development of the planet, and reason for its existence. Though it is not mandatory that you know this information it is usually easier to understand and work within an environment with which you have some familiarity.

Mars II is a planet two parsecs beyond Zeta Reticuli. The

specifications mirror those of Earth, Pylotanius, Amna, and several other planets within our allied group. Because it is positioned one hundred thousand and seventy miles farther from its sun than our own various worlds atmospheric density and planetary temperatures would not allow the development of the proper atmospheric gasses for the greenhouse effect to properly warm the planet. For this reason, fifteen centuries ago, 17,000 atmospheric processors were constructed and placed in various locations around the equator of Mars II.

These processors heated and supplied a CO_2 base for an atmosphere. With the introduction of heat, CO_2 and oxygen the construction of an ozone shield developed during the early 21st century, the planet developed liquid water in less than 100 years. Within another century aquatic plants began to emerge from the water and take root on land. An oxygen/nitrogin envelope developed slowly, and over time encompassed the entire planet. It was also realized at the time that not all future combatants would breathe the same mix of atmosphere. For that reason, nine domes more than one hundred kilometers in diameter were constructed in the southern district.

Each one had the ability to maintain atmosphere of any planet in the Warrior system. In the interest of fairness, it is stated in the rules, that combatants who do not share the same atmosphere cannot do battle on Mars II, unless it is one on one. Otherwise the expense would over ride any possible gain. At first there was debate on why it was necessary to terra-form the entire planet when the domes would do. That question was answered, when early combatants on the as yet undeveloped Mars II set up their domes and the battle began. A freak storm arose knocking out power and effectively opening the domes killing almost one million combatants, it took three hours, but they died all the same.

Although, another century would pass before another battle was tried on Mars II. It didn't take long for the administrators to realize they had a nightmare on their hands. At the time all wars on Mars II were all-out conflicts, with four or more warring factions sometimes. Details of which are not important suffice it to say a lot was learned and Mars II almost began and ended with only three wars. Over the centuries, it became necessary to restrict the amount and area of use. Before the restrictions were imposed, areas of high radiation brought about by orbital bombing with multi-megaton and plasma shock proton weapons, as well as hundreds of thousands of unburied dead, littered Mars II. The time it took for the land to reacclimatize was far more than could be spared. For those reasons the Warrior council was developed, to administer and oversee all wars, and provide rules and regulations for battle.

Eventually, Wars on such a massive scale were banned, and within two hundred years, the proper rules, regulation, and guidelines were in place with teams set up and zones allocated. The Warrior Program became official soon thereafter. The atmosphere processors on Mars II those that still worked were re-activated while the new system was being instituted. All wars of any kind were from that point on to be fought solely on Mars II, and wars anywhere else were abolished. Automated monitor drones with absolute power and impenetrable electromagnetic shields were set up to scout all planets whose representatives had signed the Mars II pact.

If any hostility of any kind was discovered the area in which it was discovered was decimated by orbital anti-proton bombing. The planet, continent, or state was simply no longer a threat. Needless to say, from that time to now, no one has been foolish enough to go against the monitors. Many variations of combatants were tried until five was considered to be the maximum number in any given war. There have of course been exceptions, as with you and your team this time. However, it must be understood that this is not the norm. Now, are there any questions before I go over the rules?"

There was silence.

"Ok; he continued "Though you have a rules list with your first briefing package, I will go over some of them now on a formal basis so I can check that off my list of obligations. They read as follows:

1.   All battles must be fought within the confines of the allocated zone. Any fighting outside the allocated zone, or within neutral boundaries will constitute a breach of regulations and be grounds for automatic disqualification.

2.   There can be no long range weapons, or weapons of mass

destruction i.e. grenades, frag bombs, incendiary devices, or missiles on Mars II. Long range is a determination of distance to destroy. In this case, any weapon that extends beyond the normal reach or fires a projectile faster than 2500 fps is considered long range. (Exceptions would be a blaster within the 50 watt range or Copchoc Nygor if velocity is less than 1600 fps. These are exceptions only because they, though deadly, can be deflected easily with Kizanti shields and most types of body armor.

3.   Teams can be no larger than five, without the express written consent of all parties. A DBY-1290/ATX must accompany any request for less or more team members.

4.   There will be no night battles.

5.   No radioactive devices, i.e. particle emitters, plasma gamma field devices, or weapons that irradiate a given area for any amount of time, with any irradiant.

6.   If a team member is killed, or wounded. They are to be left within the combat zone. When fighting has ceased for the day or moved on to another area of the zone Observation Station personnel will dispatch either a medic drone or a drone equipped with graves spray. See section 9, paragraph 178, under subheading disposition of Dead or Wounded, in the Mars II rules book. A wound which is severe enough to keep the Warrior from attending to regular battle duties, walking, running, or fighting, is the only reason a warrior is to be left behind. Examples are: Dismemberment of an extremity, blindness if permanent, loss of large volumes of blood, or a head injury that causes either disorientation or may pose a threat to the team.

7.   The treatment of prisoners of war is a non-issue. All pacts that have been attempted in the past were later found to be useless, and for the most part disregarded. Humanitarian concerns are not considered to be important, due to the customs of the various civilizations involved. To use the term human in any aspect of the rules, constitutes a prejudicial attitude, as many of the worlds concerned are not populated by beings who call themselves human.

8.   Ariel creatures, or others who fly as a regular form of

transportation, are not allowed to do so on Mars II, unless the opponents in which they are engaging in battle, have the same abilities. In which case, it must be understood, that all battles are to take place on the ground, or in the air but cannot be a combination that put the other team at a total disadvantage. Example: One team is in the air bombarding the other on the ground.

9.   No monitor, observations, or covert missions are allowed. Reconnoiter, is unnecessary, unless it is within the battle area. Here it is allowed but scanning cannot extend beyond the boundaries of the battle zone.

10.   Battle zones are appointed by the Warrior Commission and are not subject to change unless circumstances alter between the time of scheduling and the date of arrival. All battle zones are ten kilometer square. Neutral point on all four sides extending for two kilometers, govern safe areas. During battle it is not permitted to retreat to a safe zone, unless one or more of the combatants is injured. If an injured combatant is moved to a safe zone, they must remain there until such time as the referee can visually observe the extent of injury and make a ruling of validity. If ruled invalid, the injured combatant must remain within the safe zone until the battle has terminated for the day, or a victor has been determined. Beyond the safe areas, another two kilometers are set aside as base camp areas. Combatants may meet within the base camp areas, but only in a social atmosphere, not adversarial, if so desired.

There are many rules in the Mars II rules and regulations however these are the ones that are found to be violated most.

It is the suggestion of this program that you review the rules often, and allow yourself the opportunity to understand the reasoning behind each one. If the reason escapes you, there are explanatory books available. This concludes this briefing."

Computer End Recording...

Ow!...my head

It was not at all what Jarob had expected, for that matter it wasn't what anyone expected. They all kind of thought it would be a fire and brimstone briefing. You know, get out there and kick some Coonwaadii butt. You're the best team in the entire galaxy, stuff like that. But, what they got was this namby-pamby rules and regs bullshit.

"What the hell was that?" Pageena yelled as she got up and walked out the door with the others.

"Yeah...what kind of bull was that?" Jarob replied.

"I hope they don't expect us to get juiced and pumped with a briefing like that." Benni yelled.

"Hell I've seen music vids that pumped me more than that did." Marsha chimed in; "Most of them were roopa music and God I hate roopa." She continued.

"Hey...hey!! Now don't get your sphincters in a knot. That's what's known as the bullshit on the record briefing. You know the things the brass show us so they can get their appropriations for next term. You wait, blackout days are coming up fast." Campbell said.

"What the hell are blackout days?" They all asked.

"It's different every time...a kind of tradition, for warriors going off to battle. Blow off some steam you know sometimes, it's a few hours in one of the pleasure suites with the holographic girl or guy of your dreams. All the Tep you can drink for a day and a hot rag for your head in the mornin'. Usually you get a couple of days leave anyway, before going to Mars II. This time with this short a schedule, they may grant only restrictive leave or maybe none at all. Anyway blackout days, get you drunk, pumped, and stupid.

It should be happening soon...you need a few days to recoup from blackout days. Hell I have been known to make more than the usual ass of myself on many occasion of blackout days. My last blackout, me and Pooter and Jonas got really drunk and went over to station 16. If it moved we fucked it, if it was liquid we drank it, and if it gave us any shit we kicked hell out of it. God them was good times...But, Pooter and Jonas didn't come back from Mars, Brown neither..." Campbell replied.

Campbell's sentence trailed off as he stepped into the shack and headed for his room. Jarob, Pageena, Benni, Marsha and Dafora all felt a kind of giddy excitement and anticipation of what could only be one of the more enjoyable experiences they were to have on this rock.

Sleeping was hard, but they managed it. Training was as usual for the next few days, the heavy battle packs acquiring more weight with each passing step in this oppressive climate. It was odd that they seemed to weigh more each time they strapped them on, and each day they would gain a pound or two. On day seven, Hardesty came into the shack and announced that there would be a full day of briefing tomorrow so no training would take place. Campbell knew the drill everyone for that matter had some kind of idea of what was going on. That didn't in any way diminish the surprise when Hardesty appeared the next morning in a training outfit. In his hands were the training gear they had just turned in.

"I thought there wasn't gonna be any training, today?" Pageena asked.

"You thought...who the hell told you to think...or even that you could?" He barked throwing their gear down; "Get your ass into this training gear NOW! Warrior Dafora! That means full training gear with no extras!"

Quickly they picked up the training gear and rushed inside to change.

"I bet the only reason he's doin' this is he wants to see my tits." Marsha grumbled; "He's been staring at my chest ever since I arrived. Well I hope the old asshole gets a hard-on he can't get rid of."

"Oooo I'd love to find a man like that." Pageena cooed

"Oh sure...sure the old man has the hots for your tits so he's gonna run us all around the track." Jarob replied.

"Well did you hear what he told me;" Dafora whispered. "No extras. That means no bra that means I have to go out there bare chested, something I haven't done since my first day here. I bet he does just was to look at some tit to give him something to jerk off about later. Horny old fart." She continued. The others stared at her as if with new eyes.

"What?" She asked.

"Daf. Welcome aboard." Pageena said and they all laughed.
"Come on...come on we ain't got all day...move it out!" Hardesty commanded from outside the shack. In a few seconds Pageena appeared, then Marsha, Jarob, Benni, Campbell followed quickly by a shy and fast moving Dafora.

"Attention on the blue line." Hardesty commanded; "Now today were gonna run your asses into the mud, then were gonna give you two complete sessions in the ripper. But, before we begin, I have a little rumor control to take care of. It has come to my attention that some members of this team think me a horny old fart, whose sole purpose for having you get into training gear was so I could see a little tit. Maybe give me a fantasy for later. Well...well your right...Haa..ha...haaa Haaaahhhaaaa." Hardesty laughed, so hard he almost got a hernia, Campbell joined in.

"Welcome to blackout day." Campbell screeched. As both men fell to their knees in wild maniacal laughter.

"Tate you got a nice pair of melons there. Pageena I ain't never seen green boobs before...mighty nice. But, Dafora...god woman you got some huge hooters." Hardesty said obviously extremely drunk, rolling his hands around in front of him as if to fondling Dafora. Then he fell to his knees in front of Marsha and proceeded to tell her of a fantasy he had involving her, himself and a couple of Deluthian ticklers. Shortly thereafter he passed out.

"Do...don't mind him. He does this every blackout day.

He...he...he doesn't think he's successful at starting off blackout day, unless he can insult someone. Two women from Holdaria IV almost killed him last time, he did the same thing to them only then he made a grab for boobs. At least he didn't make a grab for you this time. Probably those six months in finger splints cured that." Campbell laughed.

"Ha...ha very fuckin' funny asshole." Dafora said as she covered herself and walked back into the shack.

"Yeah...you're a real scream." Pageena replied following her with Marsha, Jarob, and Benni close behind.

"O co'mon it was only a joke. We do this all the time. It's just a little joke to lighten up the mood. You know it's like the Tep in your packs." Campbell said following them.

"The what...?" Jarob commanded.

"It's kind of a tradition here, each day we put a bottle or two of Tep in your battle packs. Haven't you felt them getting heavier?" He laughed

"Yeah...but...we thought it was the atmosphere." Jarob replied.

"Well it ain't...another tradition is that you gotta drink it all." He laughed.

Jarob went to his locker and grabbed his battle pack, when he opened it all his equipment had been replaced with bottles of Tep. No less than thirty to be exact. The others were met with similar surprises when they did the same. Jarob had 30, Pageena and Marsha 32, Benni 29, Campbell 31, and Dafora had a whopping 37 bottles.

Though each bottle only held a few ounces, one bottle would usually give you a good buzz. As a side bar, it should be noted that Tep. Though created on Dafora's home world was not usually consumed by OD's. It should also be noted that you could not overdose on Tep. Even if you drank a thousand bottles your level of inebriation would only go so high and from that point on you would remain at that level for as long as it took for the Tep to wear off.

Duration of the drunk was based on ounces consumed. Weather that was hours, days or even weeks usually, it was 5 minutes per ounce consumed. However, when the Tep wore off, you had one of the worst hangovers in the known galaxy. The only good thing about the hangover was that it usually only lasted an hour and a half and that could be timed almost to the second. Then you were back to normal, whatever that might be. As a note here I should explain that Hardesty doesn't drink Tep. He hates the stuff, thinks it tastes like soap. No Hardesty was looped on good old Beefeater gin. A word of caution never… never…never mix Tep and other distilled spirits. People have been known to lose it altogether.

"May we die better than we lived." Campbell said as he twisted the cap off a Tep of his own.

"Burn 'em down." Jarob replied following Campbell's lead.

Benni, and Pageena each opened a bottle, but, Dafora was hesitant.

"Co'mon Daf, we're warriors now. We got no weaklings in this outfit." Marsha replied as she opened one of her bottles.

"Aw well…screw it." Dafora exclaimed as she twisted the cap off a bottle and began to swig it down.

"Abso-fuckin-lootly." Marsha said as they all drained their bottles and reached for another.

"'Member…onshu open tha bot-el…ya gotta dink..drink it all." Campbell slurred.

"Wha'll we drink to now?" Pageena asked.

"How bout we drink to kickin' them Waddii's ass." Jarob replied.

"Yeah." They all repeated as they again emptied their bottles.

"Noo…nowless dink to…to…to my boobs." Dafora yelled.

"To Daf's big ol' titties." Pageena replied.

"Hell we got enough Tep...we...we could drink to all your boobs...one...one at a time, an my dick too!" Campbell replied smiling broadly.

"Hell, let's get this over with." Marsha chimed in; "To all our little used, yet unforgotten sexual organs male and female."

"You got that right." Jarob replied as he emptied his bottle.

Needless to say they were all pretty well smashed within an hour of the party's opening. Dafora, toasted her boobs a few more times before she grabbed Benni, practically raped him, insisted that he bed her immediately, and afterwards they both passed out cold. Marsha and Jarob leered at each other as they heard Dafora's passion cries, then disappeared into Jarobs cube for an hour or so. Soon after cries of "Oh God." and "Yehaa." Filled the air as Marsha and Jarob shook the entire room.

Pageena and Campbell had no such inclinations however, and spent long hours in deep conversation on the ancient writings of Jean-Pierre Sartre and Kurt Vonnegut Jr. and the similarities of writing styles they seemed to share. Or at least to a couple of Tep soaked brains the similarities seemed obvious. Another thing Tep is known for is heightening the libido, a lot. Doesn't seem to work on Amnan's though, Nalvoran's either. But, it was just as well, Both Pageena and Campbell knew that the others would be more than a little uncomfortable with the actions taken tonight.

Computer End Recording...Buffer full

BEGIN TRANSMISSION

END TRANSMISSION

The Circle widens

Station 24 spun majestically in an orbit around Dulamian II. Its crew compliment of 379 from Earth, Amnan, Oberon Delta II, Xanthia, Squafo, Medaralias Prime, and many of the other represented planets of the Colonial Federation, were used to run the station. Merchants and traders however, were from all over the place. With a capacity of 23,000 it was the largest station in the known universe. For that reason it was known as Magna Station. There had never been a successful attack on Magna Station, even back when wars raged throughout this sector unchecked. The arsenal onboard was usually more than enough to counter any kind of attack; Twenty-two anti-proton guns, thirty magna-phase polarizers and another twenty or so full bore 90 inch Toka plasma cannons. Shielding up to nine levels thick and a deflection capacity of well over 85 million terrawatts per cubic centimeter.

Station 9, a close sister to Magna, had similar capabilities, but room for far less personnel. Magna station and station 9 were the only two stations in section Gamma-gamma-delta 6, or grid mark 259.47 by 340.78 in the sector known as M12. They were built roughly the same time at opposite ends of the same three parsec grid to protect the Dulamian boarders during their dispute with the Aukvon Selak. The actual war only lasted three months before they were granted Mars II space. Dulamia won and the Aukvon Selak got control of station 9, with Dulimia claiming Magna. Both stations had full hyper drive systems and could easily be moved almost anywhere in the galaxy in only a few light days. Both stations were only 17 parsecs from Mars II.

Station nine had just crossed to the night side of the small moon around Bendaiu when the first proximity alarm came on.

"Magna—this in station 9, we have a proximity on scanner 4. It appears to be in your area, but moving in our direction. Can you confirm?"

"Station 9 this is Magna—We confirm and read it as a sub-space transmission. No message can be read, it appears as thought it might be an aberrant sensor sweep that got away from someone. We cannot decode."

"Whatever it is, contact in 23 seconds, our screens won't secure it. Any ideas?"

"It's just a single source transmission, on a sub-space carrier. Shouldn't do anything but maybe ionize some of your unprotected circuits. Just to be safe you should kick on your circuit protection grid."

"Protection grids on scans indicate an active nucleus of some kind—contact in 14 seconds. 12--9--6--2----------------......"

"Station 9 this is Magna, give me your status. Station 9 repeat relay status. Station 9 we are switching to channel k delta. Get the supervisor in here."

"What is it?"

"Station 9 just went off line. It reported an anomaly in the form of a sub-space transmission. We did a sweep and found it to be a sensor sweep."

"Kind of far away for a sensor the nearest array is on Dulimia. Is that the sensor's origin?"

"We don't have an origin. But we know it came from deep space not the planet. Tracking now, point of origin appears to be M12 sector 19 around Mars II."

"That's ridiculous, Mars II hasn't got a transmitter large enough to reach even a third that distance."

"Maybe it's a bounced signal that deflected off the security field around Mars II."

"Must be but, what happened to station 9, any contact yet?"

"No sir still dead. Wait...were reading gaseous purging, and implosions. Picking up particle scatter...debris? How could there be debris if it was a simple sub-space transmission."

"Since when do sub-space transmissions wrap themselves

around a sensor sweep? Someone wanted you to think it was an errant sensor sweep. All sensors at maximum let's find out what it really was. Scan for particle emissions, ion trails, or any artificial particulate acceleration."

"We've got another one coming in...heading our way!"

"Now don't panic just scan it and keep an eye on its heading. There's no way it's gonna make it through our shields or deflection matrix."

"Impact in 35 seconds sensors read only standard scanner energy picking up traces of an active nucleus now. That's what station 9 reported just before transmission went down."

"It reads exactly like the other one, a sensor sweep encased in a sub-space field we've got a code red...secure all systems!"

"Active nucleus is beginning to pick up energy. It appears to be increasing one order of magnitude for every light-second. Sir at this rate it will have density in the neutron range by the time it makes contact. Suggest evasive action."

"Pattern Kintara now!"

"Pattern completed—object still on heading."

"Ooflek pattern and Rastara right after that."

"Ooflek pattern now initiated object will impact in 6 seconds."

Two new stars that burned brightly for an instant in cosmic time heralded the end to those great achievements that mortal beings in their egomaniacal belief once thought untouchable. No one heard the cries of the thousands, or saw the sudden flash that ended all those lives. Yet one thing was undeniable, both stations, whose destiny controlled the lives of many others, were simply gone.

Computer End Recording...

Mars out the viewport.

"HOLY SHIT!" Benni screamed.

"What the hell are you doing in my bed?" Dafora yelled.

"Your bed? This is your bed? Goddamn! What the hell have we done?" He agonized.

"We? What do you mean we? It's obvious you took advantage of me while I was in a weakened, drunken, position!" Dafora yelled back.

"Yeah? Well where I come from it takes two you know!" Benni responded.

"Hold on...hold on!" Campbell yelled; "This was a mutual arrangement. Well, maybe not."

"Ya see I told you. You took advantage of me." Dafora yelled.

"Bullshit!" Campbell replied; "If anything you almost raped Benni. Jeez you should'a seen yourself. You ordered him to feel you up first. Then you dragged him into your room and both of you were screaming like banshees a little while later. So don't be accusing him. It was all you girl...all you." Campbell continued as a broad smile crossed his face.

"I...I...I don't believe you. I...I would never do such a thing." Dafora replied.

"I didn't think you would so I made a vid-report of the entire thing." Pageena replied holding up a vid-chip.

Dafora quickly took the vid-chip, threw it to the floor and stomped it into a thousand pieces.

"S'ok." Pageena replied; "I knew you'd do that too so I got copies, and you'll never find them all."

Just then another set of blood curdling screams came from

Marsha's room. When they opened the partition, Marsha had Jarob backed up to the wall a small dagger only inches from his throat.

"What the hell are you doing?" Campbell asked.

"What does it look like? I'm gonna kill this bastard!" Marsha replied with fire in her eyes.

"Hey...hey now. Just hold on. You invited him here. Is that how you treat all the guys in your life? If it is, you may want to examine that as a reason for no return business. I mean you kill every guy that makes you and you get this real ugly reputation...you know?" Campbell replied.

"Look...Marsha...I...I'm just as mixed up about this as you are. Believe me." Jarob squeaked.

"Let him go...you wanted him as much as he wanted you. Only in this case it was more mutual than it was with Benni and Daf." Campbell continued.

"Benni and Daf...you mean B...Benni and Daf...a...a." Marsha asked.

"They surely did." Campbell laughed.

Marsha looked at Jarob, standing there with this terrified look in his eyes. Then at Benni and Daf avoiding eye contact with anyone, especially each other. Then she began to laugh, not your little tittering, giggle mind you, but a big robust belly busting laugh that came from deep down. Then Campbell and Pageena joined in, followed shortly by, Jarob. Benni and Daf finally worked up enough courage to look at each other, and within seconds they too began to laugh.

"Quite the team eh?" Campbell sobbed as tears of laughter came rolling down his cheeks.

"I guess we're starting on the next generation." Dafora replied and the laughter started all over again.

"Pageena and I were the only one's not going at it." Campbell replied; "Ain't that a bitch."

"I really don't think you would have survived." Pageena replied.

"You might just be surprised." Campbell replied.

The laughter continued for a full ten minutes before anyone realize how badly their heads hurt. When the laughs subsided, the pain set in and the horrible task of timing started.

"How long do you think we got?" Jarob asked.

"I dunno. How many bottles did you have?" Campbell asked.

"I had all of mine." Jarob replied.

"Me too." Pageena added.

"I got two left." Said Benni.

"None for me." Dafora replied.

"All mine are gone too." Said Marsha.

"Mine too. I guess that means we all got about the same amount of time give or take ten minutes or so." Campbell answered.

"What should I do with the one's I have left?" Benni asked.

"Why does the phrase target practice come to mind?" Pageena answered.

"Sounds like a plan to me...But let's wait a while." Benni replied.

About four minutes later a loud banging on the door jolted everyone.

"Tate...Warrior Tate...get your cute, but drunken ass out here. You got a priority call on my goddamn phone again." Hardesty yelled.

Marsha opened the door, looked at Hardesty's bloodshot eyes and thought for a moment to tell him to take a message. But, reason willed out, and besides she didn't think she could take more of his yelling right now. When she got to Hardesty's office, she closed his office door and picked up the head set as she turned on the vid.

"Marsha...this is a Delta red priority call. Make sure your circuit is clear and secure before we proceed." Karen ordered.

"Ok...what's up?" Marsha asked.

"Don't talk. Just listen. You and your team are going to Mars II in exactly two hours and thirty-seven minutes. That is the closest safe window. Instructions are now being sent to your commander and all associated parties. The reason for the push is that almost 13 hours ago Magna station and Station 9, her sister station, were destroyed. Both stations reported picking up a sub-space transmission before they were destroyed. Both of those transmissions were traced back to Mars II. We cannot positively identify Mars II as the point of origin.

This along with the other anomalies recently recorded only add to the urgency of the situation. Though the Coonwaadii's scheduled arrival is 73 hours behind yours, we have been granted special privilege by the Council so that we may investigate the situation a little before the battle begins.

Because of this it will be necessary to brief your team mates on the situation. This will be done in flight, with a vid chip that is now en route. This is really getting hairy now, so if you want to beg off and let another agent take over...We'll understand...I don't wanna get you killed kiddo."

"You should have thought of that before you sent me. No...no I'll stay with the team, a new guy would throw off the team's coordination. Besides when you brief them they'll have enough to contend with...Anything else?"

"No...just watch yourself...'member we still got that bottle of Tep to drink."

"Don't mention Tep to me for a while...Marsha out...see ya Karen."

The screen went blank as Marsha removed the vid-visor and laid it on the desk. In a few moments, Hardesty stuck his nose in the door and upon seeing Marsha became hostile.

"So, now the secret will be out for everyone to know!" Hardesty yelled.

"How...how do you know that?" Marsha asked.

"While you were on the phone, a sat-com came in. The vid-briefing is scheduled for in-flight and your team is preparing now. A second vid came in for me, I just reviewed it. It told me all about this secret mission and about the destruction of all those OS and ref stations. I don't mind tellin' you that I'm really pissed." He raged.

"That seems to be constant state of mind for you. Tell me Hardesty, do you plan on being an asshole all your life, or is this just an act for my benefit." Marsha replied.

"My brother, his wife, and two of his five kids are dead, dead do you hear that. Dead because you and your goddamn Warrior Program decided to keep all the shit happening on Mars II a secret. It's only because the three remaining children were visiting grandparents on Earth, that they're alive now." He yelled.

"Why? I don't understand." Marsha asked.

"My brother Paul was the Chief of engineering on Magna. His wife had just joined him there, not two days before it was destroyed." Hardesty replied, taking a large kerchief and blowing his nose. Tear were streaming down his reddened eyes as he spoke.

"I'm...I'm sorry I didn't know." Marsha whispered.

"'Course you didn't know how could you?" he replied, his voice softening in tone.

"Look." He continued, "I have nothing against you, or the

Warrior Program, it's just that somehow in the back of my mind I think that if I could have warned them..."

"There was nothing you could have done. Even if you had access to the information there was no way anyone could predict that Magna or Station 9 would be a target." Marsha consoled.

"But why? I mean what the hell is going on. What's happening to Mars II?" He cried.

"I...I don't know." Marsha replied as she laid his head softly against her breast; "But, I know one thing."

"What's that?" He asked.

"They're sending in one hell of a team to find out. I...I better get going." Marsha said.

It was not her intent to excite him, only to comfort him. Nevertheless, she feared her intent may be misjudged. Slowly she pulled away from him and began to make her way toward the office door.

"Marsha." He whispered.

"Yes sir." She quietly replied.

"You take care of Campbell and the others." replied Hardesty, staring out his window, his tears reflecting the bright afternoon sun.

"I'll do my best sir." She replied.

"I couldn't ask for more than that...god-speed." He whispered.

Marsha was so moved by Hardesty's condition, she almost broke down herself. Hardesty, even though he was such an asshole the day before, had shown a truly human side, beneath the false front of a tough, nail eating, hard ass. She didn't know why, but in some way seeing him in pain, somehow brought him closer to being human. As she walked back to shack 245, perhaps for the last time, all she could see was this grey—haired man sobbing.

When she opened the shack door, the confusion of the team's preparation was the first thing to hit her. Then everyone stopped when they saw her come in with such a sad look on her face.

"What wrong?" Pageena said as she rushed toward Marsha.

"You're gonna find out soon enough anyway. I just came from Hardesty's office. His...his brother, sister-in-law and two of their five children are all dead." She said.

"Paul...Paul and Chani are dead...What happened?" Campbell asked.

"Magna Station was destroyed last night, Station 9 too." Marsha replied.

"What...why? How? Who? Was it the Waddii's?" Jarob asked excitedly.

"They...they don't know. That's why were going a few days early. Look ah...I've got something to tell you. You're gonna hear it all in the onboard briefing but, I think it would be better if you hear it from me first." Marsha began.

"Told us what?" Jarob asked.

"I...I didn't just come here because the team needed a sixth member. I'm the reason they changed the schedule. Now don't say anything." She said as she noticed Dafora addressing herself to speak. "I'm an agent for the Warrior Program in the Office of Intergalactic Intelligence. In the past few months there have been a lot of problems on Mars II. What those problems are will be in the briefing, anything that's not covered I'll tell you about after. Initially, I was to go in with you and investigate in secret, reporting back via sat-link, if I found anything unusual. Unfortunately, recent events, including the destruction of Magna and Station 9 have rendered that program a worthless procedure everyone in the universe has heard the news by now. To be honest, I don't know why they are changing their minds and allowed you to know the true nature of the problems. I suppose they want you to know that you'll be facing something that may be far worse than the Coonwaadii. Anyway, I'm sorry I had to lie to you."

"What lie? You didn't lie to us." Jarob said.

"A lie of omission is still a lie." Marsha replied.

"Yeah...well, let's face it, all of us had some idea that something was going on...right?" Benni said.

"Yeah." They all replied in unison.

"It's kind of foolish of you to believe we didn't suspect anything. I mean a sudden six pack team, you knowing everyone and their brother who works here. We all know the security clearance that requires. I didn't want to say anything, but I had a friend of mine pull your file. Most of it was classified Ultra, meaning up until a few years ago you didn't exist. What I found out most of all is that for the past few years you've been attached to the Warrior Program under Karen Turbin of OII. A couple of years ago you got hurt real bad or something...no need for detail. Suffice it to say what happened soured you on field work. So until you came here you were a desk jockey." Campbell said.

"That's about right. You might say I got the job because I was available, already had the necessary training and most of all I'm expendable, just like the rest of you." Marsha replied.

"You know I thought I had heard of you before you invented the scanner we're using...Right?" Dafora asked.

"Among many...that was about the third I guess." Marsha answered.

"You're very well known on Oberon Delta. At least in name, some of your theories on scanning waveguides are still being used. I recognized the name immediately, but I was hesitant to ask." Dafora replied.

"Why?" asked Marsha.

"Well, to ask would mean I didn't know, since OD's think

everyone inferior to us. I couldn't bring myself to admit that. I'm afraid my social upbringing got in the way again." Dafora replied.

"You sure have come a long way in the past weeks." Benni said.

"Anyway, that's about all I have to say for now. I just didn't want it to come as a surprise to you when you heard the briefing." Marsha confessed.

"S'ok, forget it. Now let's get your battle gear packed and get over to the shuttle bay. You might want to get into your dress uniform too." said Campbell.

"Let's go kick some Waddii butt." Jarob yelled.

"Yeah...yeah...yeah." They all yelled as they tore out of the shack and made their way full bore to the shuttle bay.

"Y'all know the drill. Just find your assigned rooms and stow your gear." Campbell said.

"What? This shuttle is identical to the shack. What's the deal Campbell?" Jarob asked.

"The pod we're in is the exact same layout as our rooms in the shack for a lot of reasons. The best one being that, when we arrive at Mars II, this entire pod will be dropped and piloted to a predetermined location. Since the old shack has been our home for so long. We will feel less disorientation which will in turn make acclimation easier. If necessary we can move the whole thing to an alternate location using the auxiliary shuttle controls, in the forward hatch there by the view screen. Though we could achieve orbit, and our fuel supply is good for many light years, it was designed for short jumps only." Campbell instructed.

"I guess we should do our pre-flights." Pageena said.

"All the gear is stowed. Raise the launch station." Campbell ordered.

Mars II

The launch station rose in the middle of the main room. It consisted of six seats, with high backs and the necessary belts to strap them in. In front of each chair was a control console.

"Launch in twelve minutes." Came a voice over the com line.

Computer End Recording...

Stolen moments in the darkness

"Pageena you take op's, Marsha and Dafora your tech stations are on either side of her. As for Jarob, Benni and I, just find a place and park it." Campbell said.

Jarob and Benni took the two seats behind Marsha and Dafora, with Pageena and Campbell taking the two behind them.

"Hey! I didn't think you was boss anymore." Jarob teased Campbell.

"Just the voice of experience my young friend...the voice of experience." Campbell replied with his usual hearty laugh.

They all grinned and waited as the countdown began. When the green launch indicator came on and the main turbines kicked in, the shuttle surged forward as if shot from a cannon.

"That wasn't so bad." Jarob said of the remarkably smooth takeoff.

"Yeah...not too bad at allll...." Benni replied as they hit the first security barrier and bounced around roughly for a few seconds.

"We still got three more security fields to get through." Campbell said tightening his seat harness as they struck another.

"Why don't they turn the damn things off." Pageena asked.

"It's a restrictive field. Seamless that means planet wide. Kill one part of the field you'd have to kill the whole grid. So...so they put extra shielding around outbound ships and punch them through. The field is self regenerating so it's easier that way." Campbell said as they bounced, banged and boomed through the next three security fields.

"That's...that's it." Campbell signed.

"Pageena what's our E.T.A." Jarob asked.

"E.T.A. is fourteen hours, nine minutes and thirty-four

seconds."

"We have a green on all control systems, and our course is steady. Apart from a minor course correction we have to make in about three hours, we confirm Pageena's E.T.A." Marsha said looking at Dafora.

"Before you guys get up we have a few internal systems checks to make." Pageena said.

"Like grav-systems, right?" Jarob said as he noticed the diepoit disk in his left uniform pocket float out and up to the ceiling.

"Right." Pageena replied with a smile.

"Ready ladies." She asked.

"On line now." Dafora replied.

"Coming up to syncro...now." Marsha replied.

"Ok systems running green, we have grav, O2, lighting, heating and all secondary redundant. Ok, life support reads green. Do I have a confirmation from tech station?"

"I don't know about tech station, but I'll confirm that." Jarob replied as the diepoit disk came back down and cut a two inch slit in the bottom of his pocket.

"We confirm that." Marsha replied.

"Ok all automations are on and all systems seem to be up and running." Pageena replied.

"Guess that means it's briefing time." Campbell said as he opened his chair harness, stood up and began to walk toward the vid screen. Jarob, Marsha and Dafora moved slowly into the vid area with Benni and Pageena close behind. Campbell then opened the envelope containing the vid-chip, placed it in the scanner and they waited.

The following briefing is of highest security, before

continuing, the internal security network contained within this chip will scan your systems for any signs of intruders. Scanning, scanning, scanning....scan completed system secure. All warrior team members will step forward for retinal scan. Scanning Timbo, Pageena...cleared. Scanning Ka, Jarob...cleared. So it went until all had been cleared by the damn thing.

Then Karen appeared and the briefing began in ernest:

A little more than four months ago, when the Baldoris and Mmal Ju were to battle on Mars II, both sides disappeared. Since that time no less that twelve other parties have also vanished. The latest being the Ypsilanti and Bohabak, nine days ago. The Bohabak, did not answer the morning call to arms. As is usual in that event an observation station was called in to review the last known coordinates of the Bohabak. Exactly three minutes later, the OS was destroyed with a surface to air missile, which we all know are not allowed on Mars II. In each case, an OS was either damaged or destroyed in a similar manner.

For that reason we have requested and been granted, for this event alone, a team size upgrade, to six. We made this request in order to allow one of our operatives, Commander Marsha Tate, to investigate further, by becoming a member of a team in training, then investigating covertly, once on Mars II. Now in retrospect we see that an error was made in this particular plan, because it was not timely enough. We waited almost four weeks, and we could have had someone on Mars II in that time. But, to be honest, the Warrior Program hasn't allowed us to place anyone but warriors on Mars II, according to rule 285, paragraph C, subsection 9, so we assumed that our initial plan would be the best bet. However, the events of the past eighteen hours have forced us to again push the time table forward.

At 18:30 hours yesterday, station 9 was hit with an energy force 15 times greater that her shield could dissipate at maximum. Six seconds later she exploded, killing all 2,567 onboard. Five minutes later and identical force smashed the shields and deflector of Magna station, where close to 23,000 were killed. Both beams seemed to have Mars II as their point of origin. Though it is not known how such a beam could be fired from there undetected. Indeed we cannot be absolutely sure that Mars II is the point of origin and not just a deflection point. However, evidence is strong enough to warrant the immediate investigation of Mars II. Since your team was the next in rotation, we decided to attach our operative. But, as of now you are all operatives, like it or not.

Your mission is to investigate the circumstances on Mars II and report back any and all anomalies. Commander Tate has a transceiver that is capable of sat-link and is authorized to use the link five times. That authorization now extends to you. Since you will be arriving 73 hours before your battle time, you have that much time to get to the truth. If by the end of that time no information can be found, the battle will go on as usual. Since it is impossible to know the final outcome of that battle, let us hope it is in your favor, the victor in either case will receive the assignment and the investigation will continue. Despite any and all failures we will get to the truth, no matter how long it takes. We here at Warrior Central bid you luck in your battle, and good fortune in your investigation. Each surviving team member will receive and extra chit for their part in this investigation. Thank you. This is Turbin, Office of Intergalactic Intelligence. JFR-1005979-5611 #00197 red-code-red-yellow-blue. File #109 BZET Alpha 919 END UPLINK---END TRANSMISSION.

"Commander Tate...I'm impressed." Pageena mocked.

"Yeah, well don't be...I didn't even know I was a commander until just now." Marsha mocked back.

"Let me see if I got this right? We're gonna go to Mars II to investigate the downing of OS units and the disappearance of warriors missing from past aborted battles?" Dafora asked.

"That's what the lady said." Jarob replied.

"But, after we investigate we're suppose to kick the Waddii's butt, then go on investigating. Right?" Dafora asked.

"Yep." Marsha replied.

"Well, it's clear as mud now." Dafora replied as she stood up and walked toward her cubicle.

"The good news is, if I get through this, I rotate out. I only need two more chits to get out. That means I'm working on short time now...He..he...heeeee." Campbell replied.

"Hey that is right...alright Campbell!" Jarob yelled.

"Hey Marsha is there anything else we should know?" Jarob asked.

"No. No I guess not. She told you everything I've been briefed on. 'Course you got to remember who were dealing with. It's more than possible that we were given only the information they wanted us to have." Marsha replied.

"Did you know this stuff before you got to Triton?" Dafora asked.

"Some of it but, most of the time I had to rush over to Hardesty's office and pick up the latest." Marsha answered.

"I think that's when we first started to get the idea that all was not what it seemed." Pageena added.

"I can still hear Hardesty yelling about all the cloak and dagger stuff." Marsha replied.

"Yeah, Hardesty never was the most patient guy around." said Campbell.

"He certainly didn't like me." Marsha said.

"Like you...The man would have bedded you in a heartbeat. The scruffy ol' letcher." Campbell added.

"He never gave me that impression." Marsha laughed.

"Bullshit! blackout days should have told you that." Jarob said.

"Oh co'mon he was drunk. He wanted to have a little fun."

Pageena said.

"Some fun? Making us put on our training gear, and I couldn't even wear my new bra." Dafora smiled.

"It's happened in reverse. Commander Stillwell, a female commander mind you, once made all the male members of all the teams do a line up naked. She said she was checking for lice. There ain't never been lice on Triton, ever. Not ever. Then she'd have cock parties and dick wrestling matches. She was a real piece of work. But, whoa booka if she ever got drunk. Wheshu! The lady was a real animal." Campbell confessed.

"She ever get you Campbell?" Pageena asked with a gleam in her eye as Marsha and Dafora waited anxiously for his answer.

"Well I...I." Campbell shuddered.

"Yes...yes. I knew you were a hound at heart." Marsha teased.

"Hey all this happened before I met my wife. Stillwell's been dead almost 246 years. I was a rookie, you know. An for an older woman she was fine." Campbell said trying to defend his honor.

"You ah. You ever make it with a recruit?" Jarob asked.

"Now what kind of question is that? I know better than to mix up with some recruit..." Campbell replied indignantly. "...hell I almost did once." Campbell said with the appearance of anger first, then the gentle tone of a man tempted almost beyond his limit.

"I think the conversation has degraded. I'm gonna try to get some sleep before we land." Marsha said.

"Good idea." Campbell said obviously glad the conversation was over. "Jarob, Benni, Pageena, Daf, you comin'?"

"Nah...I'm to keyed to sleep. Maybe I'll just check the vid, watch an old movie or something." Jarob replied.

"Yeah...sounds good to me to." Pageena replied.

"Sleep sounds good to me." Benni said.

"Me to." Dafora replied.

"Your bed or mine." Benni asked looking at Daf.

"Not tonight, I have a headache." Dafora replied with a smile.

Pageena and Jarob sat in silence for a while, flipping through the vid-files. Both finally settled on a remake of the classic Marook Obcomin. A sensitive tale of alien meets alien, alien falls in love with alien, alien loses alien tender snore inducing stuff. They watched it for about a half hour then realized they were moving closer to each other. Soon a warm embrace, and some time later Jarob woke up in Pageena's bed. Both were happy for the experience.

Jarob had not felt happiness, real happiness, in a very long time. But, this time it was different. The sharing, the knowing, and when they had done, Pageena was at that point and forever, a part of him, his life and every facet of his existence. The sex was good, but that wasn't all there was to this. In a brief time, he knew all about her, her race, their struggles, her father, her mother, her brothers and sisters. Your entire family going back for hundreds of generations.

But, it wasn't just, knowing someone. It was as though he had lived each life, had loved each one, as if each experience was his own each life only a single moment in time. He was born, aged and died all in the span of a few seconds. Yet at the end of each life, he remembered, and retained all the knowledge accumulated through that life.

"That was the most intense thing I've ever felt." Jarob said in amazement.

"It was short, but it was interesting." Pageena replied.

"I know you. Your whole family…even your pets. I even know what you were like as a child." He said.

"But you, your life was so violent, so sad. The anger, the hatred all I seemed to get from you was violence. There were a few brief moments of love, when your mother was alive but that Dwayne, what an asshole." She said.

"Yeah I know. He's the central reason I left home when I was a kid." Jarob said.

"I could see that. But, there have been many great achievements your family is responsible for. Barooth Ka your great grandfather established the first Kaldite settlement on Caldoromus III. Your great grandmother was killed in a rock slide shortly after they had settled...So much sadness." Pageena said.

"I...I never knew any members of my fathers family, he died when I was young." Jarob sighed.

"Yes...I know. But, your family has a long history of honor, wisdom and love." Pageena said.

"I could see, and feel the love and security your family had. In fact all of your people had the same kind of deep, secure, loving. I have to admit it was great to experience it. But, I think I can understand now, what you meant by your emotions lacking substance. In the midst of all that love, all that security, all the warm, good, feelings, there was this emptiness. Not anything that you could really put your finger on, just a kind of blackness that hovered around." Jarob explained.

"That's the HoMas, the dead place. Part of our growing, is knowing and exploring that place. My father used to say, if you can see it, feel it, and smell it, what good is it if you can't eat it?" Pageena said.

"I felt your love for him. He died last year didn't he?" Jarob asked.

"Yes..." Pageena replied her eyes welling with tears.

"Well...;" Jarob said putting his arm around her and pulling her close; "You've got him up here, in your head don't you? Now so have I. So I guess that means he ain't really dead right? In fact he never will die as long as your line continues, and I have a feeling that's gonna be around for a long, long time." Jarob consoled.

Jarob and Pageena drifted off to sleep in each other's arms as the craft sped slowly, yet steadily toward an unknown destiny. In a couple of hours the bustle of the others roused him, as they made the necessary corrections. He looked down at the lovely face of the sleeping Pageena, and then out the viewport. Mars II now stood out starkly in the darkness of space. Pageena woke and looked at him. Gently he brushed her long black hair from her face and kissed her gently. They held each other, and gazed for a few silent moments at the blackness and the planet that held the answers for all of them.

Computer End Recording...

Mysteries.

Mars II was not at all the planet one might assume it to be. The original Mars was a red, barren, wasteland until it was terra-formed. Now even though it has running water, plants, and animals. Its red color still saturates everything. Mars II on the other hand is blue, green and yellow. Rich with the colors of ochre, and deep dark browns, with a ratio of water to land almost exactly 50 percent. The leaves change with the seasons, and the water is as blue as the seas of Earth.

Beautiful...yes... But, there are some things one must get use to. The pressure for one, made almost all the trees grow squat and wide. The heat was bad, especially in the summer months, which as a point of information, I should say, it is now. Because its size is roughly three quarters that of Earth, the gravitational density necessary to sustain life, at least life as we know it was too much for the planet. Thus the pressure rose and the heat from that pressure could not bleed off into space, due to the EM field set up by the atmospheric processors. Oh Mars II was a beautiful place. But as with anything else you sometimes have to compromise to get the beauty you want.

"Shuttle is now in upper atmosphere. Prepare for drop in 15 seconds." Pageena said as she and the other sat tightly strapped into their chairs.

The shuttle they were using was a drop ship. It would drop the pod with their housing, and other equipment, set up a magnetic flux variant to allow a controlled soft landing. Then take up a station keeping position in orbit until it was called for a pickup.

"Mag-stat's on line...3-2-1 release." Dafora counted.

When the release valve opened the pod began to drop like a stone until the variant was equated by the computer, say two—maybe three hundred feet. At any rate it felt like they had fallen a few thousand, and genuine panic began to sweep in until the controls came on line.

"Wasn't that fun?" Pageena said sarcastically.

"Always was for me." Campbell replied.

"Benni you can open your eyes and let go of my hand now." Dafora said trying to wrest her hand from his crushing grip.

"Oh sorry." He replied releasing her.

"Do we have a coordinate lock?" Pageena asked.

"Telemetry coming in now…should be about 250k over that rise to the east." Marsha replied

"Were tracking by grid glide angle is being fed to the nav-com. Syncro-lock in 12 seconds." Marsha said.

"Passing grid 149." Pageena reported.

"9 seconds." Marsha counted.

"Grid 126." Pageena continued.

"5 seconds." Marsha continued.

"Grid 111." Pageena again reported.

"We have syncro-lock on nav-com at grid 93. On station in 6 minutes 47.031 seconds, set landing cycle on my mark. 4...3...2...1...mark." Pageena instructed.

"Landing cycle engaged, landing protocol shows green. Drive thrusters and mag-flux are cutting to zero." Dafora replied.

"Kind of cool having two techs, sure makes landing this thing easier." Pageena said.

"You got that right. Most of the time the auto-sequencer, doesn't engage in time and the landing get really rough." Campbell said.

"Well we're not even using the damn thing, so this should be a piece of cake...Wait a minute...maybe not...picking up ionization in the primary guidance pod. Might have stripped a shield switching to redundant systems...there...there that seems to have cleared the problem...beginning descent." Dafora said calmly.

"Approaching pod this is referee 211 checking com line."

"Referee 211 this is pod 16 from Triton, do we have clearance for sector 09, grid 31, area 43?" Pageena asked.

"Checking now...clearance granted for sector 09, grid 31, area 43. Upon landing you are to contact Battle Control on frequency K-delta-2-7-0, designation Warrior 6. Good luck and Welcome to Mars II. Referee 211...out."

"Copy...and will comply...Warrior 6 out." Pageena replied.

The large pod drifted slowly downward as each second passed. Actually, it would be more appropriate to say pods. The unit itself more closely resembled a cluster of pods, with four making up its base, and another sitting top center. Two of the four base pods were for housing, one was for battle prep, with the last being control con, communications and op's. The upper pod was a storage and instrument package. It could also be detached to double as an escape pod in the event of an emergency. It was the only piece of equipment that was not a regular part of the usual Warrior complement. In normal times the cluster of four was more than enough.

"We're drifting left a little, compensating." Dafora said.

"Target zone is locked and tracking to zero." Pageena replied.

"Maneuver completed...touchdown may be a bit of a bump. We caught a bit of a downdraft off that mountain range to the west." Dafora said.

"How fast?" Pageena asked.

"Oh about 35 kph. The stabilizers should hold us...contact in 38 seconds." Marsha said.

"Hey Pageena why the hell are you flying this thing? Aren't

you tactical?" Jarob asked.

"Yeah. But, I'm also the only pilot here rated on this design." Pageena replied.

"What about Campbell?" Benni asked.

"Hell...I been on these damn things. But, flyin' em? No thank you...I have a hard enough time drivin' a mag-car." Campbell said.

"How'd you get a rating for a warrior design?" Jarob asked.

"I'm rated on the control console, not the ship. The ship could be any configuration so long as the console design is the same. Now shut up and check your harness." Pageena replied.

With all harnesses checked, white knuckle time began as they descended closer and closer to the target zone. Bump-bump, rattle-rattle, shimmy-shimmy-bang, and they had landed.

"Thrusters off, secondary and primary fuel pumping systems off line. Shut down com-nav, and all redundants." Pageena ordered.

"All systems down and secure." Dafora complied.

"Well here we are...what now?" Jarob asked.

"We only got about two hours of daylight. Maybe three. You think we should do a little scouting?" Campbell asked.

"Maybe, what is the first set of grid coordinates we gotta look at?" Marsha asked.

"That would be 19, 29, 42, about 3k east. Let's see, according to the map that would put us in the foothills of the Cobi Cori mountain range, about 2k off the battle zone." Dafora replied.

"We could reach that by mag-car in about ten minutes." Jarob added.

"Let's go." Benni said.

"Now just hold on." Said Campbell, trying to be the voice of reason.

"What?" Benni asked.

"Are we allowed to go into that sector armed?" Campbell asked.

"I...I don't know." Benni answered.

"Don't you think it would be wise to find out?" Jarob asked.

"Yes...I guess so." Benni replied.

"Pageena...would you call Battle con and find out what the restrictions are for that area?" Campbell asked.

"I got them now. We had to call them anyway. Battle con, we have a query about 19, 29, 42, 3k east, what is the arming status?" Pageena answered.

"Warrior 6, you are restricted to short range standard in that area. Plasma charge level .30 or below."

"Plasma charge level .30, we copy...Warrior 6 out." Pageena replied.

"Plasma level .30? That's just barely above medium stun. Are they kidding?" Campbell yelled.

"No that's the rule, and you know violation could disqualify us if we break it, even if we are not in battle." Pageena replied.

"Jeez, so what are we suppose to do if we run into trouble. A plasma .30 would barely be a bee sting to some of these critters around here?" Jarob asked.

"Wait he said short range standard, that means diepoits too right?" Benni asked.

"He did say short range standard." Pageena answered.

"Call them back and confirm short range standard, make sure that includes diepoits." Campbell ordered.

"Hey...you ain't the boss here anymore." Pageena yelled.

"Hold it...hold it, this isn't the time for tempers." Jarob replied.

"Yeah...sorry, Pageena." Campbell said.

"Me too...I'll call in and confirm diepoits. Why don't the rest of you get ready?" Pageena said.

While the rest of the team got their gear. Pageena, called Battle con and got the ok to use diepoits, but only within a 50 meter range.

"We're cleared." She told the others as she slipped on her battle pack and got into the mag-car with the others.

Campbell was at the controls, which didn't sit to comfortably with the others after what he had said earlier. But, he proved to be much better than he had give himself credit for.

"Coordinates check." Dafora said, looking at her scanner.

"I read E.T.A. three minutes." Campbell replied.

"Confirmed." Marsha said.

"Confirmed." Dafora repeated.

"Should be just around that small knoll up ahead." Pageena said pointing to a small outcrop of tall grasses and stones.

"Wait a minute, I'm picking up trace particles of some kind." Dafora said.

"Me too. Reads like some kind of pyrotechnic trail. Definitely, a fuel/air combustion around here some time in the past few days. Look at this Daf, my readings go right off the scale." Marsha answered.

"Mine too." Dafora replied.

"Could it be residue from an OS or a referee platform?" Jarob asked.

"No way...that would be an ionized axiom emission. This reads like a carbon combustion of some kind, like an old style solid rocket propellant." Marsha replied.

"I'm reading a great deal of unspent fuel. Very inefficient, probably a fast burning, low heat, exchange of gasses, unless the rocket was as big as one of our pods. I'd say its range wouldn't exceed...say...12k." Dafora added.

"That would be my guess too." Marsha agreed.

"Wait a minute...You're telling us that some one fired an old style missile from here?" Benni asked.

"That's the indication." Pageena answered.

They were all a little jumpy as they walked around the large outcrop of jagged rocks a few meters off the edge of the knoll.

"Look here!" Dafora yelled almost scaring the others to death.

"What?" Asked Marsha tersely.

"There's a blast pattern on the rock face...see?" Dafora asked.

"No I don't see." Benni replied trying to recover from the fright.

"Me neither." Jarob agreed.

"Look...look here, see this pattern that spreads in a rosette. Someone went to a lot of trouble to disguise it but if I take a reading at the center. See...the center is fused as if subjected to a great deal of heat." Dafora explained.

"Yeah...yeah I see it now. She's right." Jarob agreed.

"But what does all that mean?" Campbell asked.

"It means that whoever fired the missile was standing about here when they did it with their back to that rock." Dafora replied.

"Look here there are still residual traces of unspent fuel blasted into the rock. Judging by the decay rate, it's probably a week or more old." Marsha added.

"That doesn't make sense...look around...look around on the ground. The soil is sandy here footprints should be all over this area. See...nothing." Jarob said.

"Has there been any rain here in the past week?" Campbell asked.

"No...the last rain that fell here was more than a month ago." Dafora replied.

"But, this is a blast point. I'm positive, that this is where a missile was fired. These readings confirm that." Dafora said with Marsha in quick agreement.

"You pick up any life signs besides us?" Jarob asked.

"No...no just us, the ref station about twelve k east. Maybe if I boost the range..." Marsha replied; "Nope, my range is maxed out and the only life signs are the ones who are suppose to be here."

"I'm picking up what looks like a synthesized energy field, about twenty or so k in that direction." Dafora added pointing north.

"What band are you in?" Marsha asked.

"R-delta." was the reply.

"Switching now...yeah there it is...definitely a synthesized pattern. Might be an emission from one of the atmospheric processors?" Marsha confirmed.

"I don't think so...the map doesn't confirm and AP in that area.

Maybe there might have been one there at one time, that got destroyed during an ancient war. If the taps were still on line it could generate a synthesized signature." said Dafora.

"But, the decay on those taps would have died out years ago. According to ancient rules, AP's couldn't be targeted if one was destroyed it was an immediate disqualification." Benni said.

"Still, it's possible that the readings we're getting are from either an AP or perhaps some other equipment that may have a regenerating power supply." Campbell added.

"I don't think so. Any equipment left behind was to be collected under contract by the Spacing Guild. Those contracts were huge, and covered more than 1200 pages, and stated specifically, that if there was any equipment left on the planet after the salvage effort, the contract would be voided and no payment would be made. So, you can be sure that after each war the zones were combed thoroughly." Pageena replied.

"Maybe these were left before the contract." Jarob said.

"What? This contract has been in effect for more than five centuries. No known regenerative power supply could last that long. Even regenerative systems sooner or later run out of raw materials." Pageena argued.

"Could it be residual radiation?" Campbell asked.

"No the particle emissions are too unstructured. It has to be a random source discharge. Like a generator or something." Marsha said.

"Well, I say we take all these readings back to base and see if we can get a pattern of some kind out of the computer." Jarob said.

"I agree." Pageena replied.

"But, with the limited time we have before our battle begins, don't you think we should investigate the area of the power emission?" Dafora asked.

"We can launch a VR probe from base to scout that area. Besides by the time we go there it would be dark, and we couldn't see a thing anyway. The VR has infra-red and ultra-violet tracking onboard, we'd get a better readings that way anyway." Jarob answered.

"But, isn't that illegal? I mean to fire a probe this close to battle time." Pageena asked.

"No...I don't think so...besides its scouting point would be outside our battle zone by more than 20k, and the Waddii's ain't even here yet." Campbell replied.

"Just to be sure, we should get an ok from Battle Control." Marsha said.

"OK...let's finish our readings and head back." Jarob suggested.

Base camp was located at the foot of the northern slope of the Cobi Cori mountains. From the viewport one could see a vast flat plain now aglow with the colors of a sunset fire. Three of Mars II's four moons now rose in majestic procession in the east, adding new light and dimension to the darkening horizon. Hopping Japro moved swiftly and silently to their night feeding places and dozens of small rat-like Hmook scurried to get out of the way. Evening also brought out the Kyzalian passion birds singing their high pitched calls, followed by low cooing.

The team however took little notice of these spectacular event as they returned and began to carefully review the data they had gathered. After inputting all information into the system, they had to wait a few hours for the computer to research, compile and download to the mainframe at Battle Central. Since they had Battle Central on line anyway, they asked about the probe and received the go ahead.

"Probe launched." Pageena said.

"Receiving telemetry, and VR link information...Link established." Dafora replied; "Marsha you'd better take over, I'm not qualified to pilot a reality probe."

"I'm not either...go ahead." Marsha replied.

"Coordinate lock...IR and UV on and reading...picking up trace patterns...IR reading heat trails, seems to be something moving." Dafora said.

"Is it alive?" Marsha asked.

"Can't tell...doesn't look like Jopro, Hmook. Still, it could be some kind of animal. Switching to UV, there...that's better. It...it's a transport of some kind...UV reads three...no five life forms inside. Strange, the transport reads as alive too."

"Could be Waddi?" Campbell replied.

"No...no the body signature's all wrong. Can't tell...looks kind of like humanoid, but...Hell I can't tell. Wait...wait it's stopping. They're getting out...I...I think they've spotted the probe."

"Get it outta there." Jarob yelled.

"Maneuvering now...switching back to IR. I've got indications of four other vehicles approaching from the east...There's a flash. Tracking indicates the probe has been fired upon...taking evasive action."

"Who the hell is shooting at our probe?" Benni asked.

"I got a better question. Where the hell are the monitors that are suppose to watch this place when no wars are going on?" Marsha asked.

"Come to think of it, where were the monitors that were suppose to be watching Station 9 and Magna. Think...think have any of you seen a monitor around?" Jarob asked.

"No...no, but that's not unusual for here. Mars II is the only place that does not require monitor control on a regular basis." Pageena replied.

"When there's a war going on, it doesn't. But, any other time Mars II is supposed to be monitored just like any other system. We've been here hours and we haven't seen a single monitor in all that time. Usually, you'd see one or two." Marsha said.

"I've got a missile lock! Trying to evade...cliffs ahead, I'm going to try to lose the damn thing...if I time it out right, I should move just fast enough to..."

"Don't talk, do!" Marsha interrupted.

As the probe swung wide around the first peak of the tall black cliffs a few k from the target sector. The missile, in hot pursuit, began to gain ground. A small slit between two cliff faces appeared in the view of the reality probe. Above it and to the left a small jagged shelf hung precariously. The probe then thrust forward through the slit and then began a long high loop. The missile followed true to its program. As the probe reached the bottom of the loop. It made its way directly toward the slit again, and the small shelf above it. The probe passed within one meter of the shelf, creating sufficient backwash off its propulsion to dislodge it. As it fell, the missile came into direct collision. KABOOOM.

"Daf that was some pretty fancy work. You sure you've never flown a VR probe before?" Marsha jibbed.

"Can you take over now Marsha...I...I'm burnt on this. I've got it on auto for now it should orbit till we cut it off." Dafora replied.

"No problem." Marsha said taking the control visor and gloves.

Computer End Recording...Buffer full...

TRANSMITTING

END TRANSMISSION.

Who fired that?

Marsha continued to remote the probe until it had been safely retrieved and replaced in the launch bay of the upper pod. A small disk onboard used as a backup information system, was retrieved and taken to ops for further analysis.

"Let's run a parallel scan using both the backup and the recorded information we got from the probe direct. Maybe there's something we missed." Marsha suggested.

"Backup is on-line now. Setup is ready for syncro-scan readings coming in now." Dafora replied.

"There's our first contact with the transport. Looks like two humanoid figures got out." Pageena added.

"Can we lock and zoom in on that area?" Jarob asked.

"Vid-mapping…on. There." Dafora replied.

"I can't see anything." Campbell said.

"Neither can I.' Jarob agreed.

"I'll try to enhance, but it's pretty dark out and even with the IR and UV there's a good chance we won't get much anyway." Marsha replied.

"What about if we mix both signals, wash them a few times, then feed them to the enhancer?" Pageena asked.

"How many washes?" Dafora asked.

" Don't know...three?" Pageena suggested.

"Ok we've got a frame lock overlaying now. That does seem to bring out a bit more detail. The wash and enhance will take time though, there's an awful lot of trash in the mix." Marsha said.

"How long?" Campbell asked.

"Four, maybe five hours." Dafora replied.

"Well, if it's gonna take that long we might as well check equipment, maybe get some sleep." Benni said looking at Dafora.

"I'm not really sleepy this is set for auto think I'll check my stuff." Dafora replied giving Benni a knowing smile.

"Well, I'm gonna get some shut-eye." Jarob said.

"Me too." Campbell agreed.

"Sounds good to me too." Pageena said.

"I'll watch the console for a while. Maybe download our findings to the Battle Control computer. See what they make of it, sleep well." Marsha replied.

It didn't take long for the clatter and hum of the equipment check to give way to silence, as they completed their checks and were soon in the warm embrace of sleep. Marsha sat at the com system and watched each line as the scan began to trace down the screen. Each scan brought out minute changes in the overall picture. One line ended another began as slowly, hypnotically, the computer traced each image again and again.

A smile crossed her face as she heard the faint footfalls of someone going into Dafora's cubicle. Then Daf's slight giggle she had come to recognize in the short time Marsha had known her. A few moments later and the giggles turned into subdued passionate cries. Within an hour however, again the sound of silent footfalls and again there was silence. Blip---hum---blip blip, the computer sounded as each cycle was completed, then repeated.

"Warrior 6 this is Battle Control. Come in Warrior 6 over." Blasted into Marsha's headset shaking her awake.

"Um...Yeah...um Battle Control this is Warrior 6. We copy over." She replied.

"Warrior 6, we have a confirmation on your downlink. It looks like someone out there has a good supply of modified, visual tracking, surface to air missiles, they're ancient of course, and may or may not work. But, indications are positive on a missile track...over."

"No shit...tell us something we don't know, over."

"How 'bout this. The monitors are gone, over."

"Yeah we know that too, over."

"No I mean all the monitors are gone...everywhere, over."

"You mean just around Mars II right, over?"

"No, I mean according to reports we're getting it looks like every monitor in the entire sector is gone. Maybe all of them in the entire system, over."

"Any word on where or why? Over."

"Nope, we just know their gone, over."

"Keep us posted. Warrior 6 out."

"Will do. Battle Control out."

At first Marsha wanted to wake the others and tell them. But, after thinking about it, it wouldn't make any sense to them either. So, she figured it would be best to give that information to them in the morning when the image enhancement was completed. It was possible, however unlikely, that the image in process might shed light on both mysteries. Marsha, stood, stretched, yawned and headed toward her cubicle and the soft, warm bed that now seemed to call her name in silent voices. As she lay down the blip---hum---blip blip could be heard from the main console, soon the rhythmic patterns of the computer chatter, eased her off to sleep.

Though it seemed but seconds, hours had passed. At 05:45 standard Earth time, the warning klaxons went off waking everyone with its loud clanging and low pitched whoops. The red flashing lights and bright blue alert beacons only added to the confusion of waking with a start, and then dealing with whatever as it happened.

Jarob rushed half-naked into the op's area with Benni close behind, blaster in hand. Campbell rushed out wearing a t-shirt, and boxer shorts with little red dragons on them. A cigar dangled out of the left side of his mouth, creating a puff of white smoke that drifted around his head. His eyes were wide, with a wild look in them. With a mobar in his right hand and his plasma rifle in his left, he was sure he was ready for anything. Dafora and Pageena came out at the same time and immediately went to their stations. Marsha ran out of her room and almost fell over Benni who had taken that moment to pull up his socks.

"We've got a red light on the com system." Pageena yelled above the din.

"What's that mean?" Jarob yelled in reply.

"Checking now." She replied; "Warrior 6 to Battle Control. What is the nature of your code red? Over."

"Code red is six-niner-zero, repeat six-niner-zero, we... explosions in all.......and our backup systems ar..beginning...fail. All battles are suspen...until…" the static reply began to fade to white noise.

"Battle Control! This is Warrior 6, do you copy? Battle Control? Battle Control do you copy?" Pageena repeated over and over in an urgent plea as the signal faded.

"What's going on?" Dafora yelled.

"Check the code for six-niner-zero. Battle Control just went off line and I can't raise them anymore." Pageena commanded.

"Six-niner-zero is Battle Control under attack!" Dafora replied in a moment or two.

"Under attack? Campbell asked; "By who?"

"Unknown. But, apparently they were caught off guard."

Pageena replied.

"Where are the monitors? They should be handling this." Jarob added.

"Their gone...they're all gone." Marsha replied; "When Battle Control called in last night to confirm our probe findings, they told me that all the monitors were gone. Not just here on Mars II but the whole system, and maybe the entire sector."

"Gone where?" Benni asked.

"I couldn't tell you. All they said was that they're gone." Marsha replied.

"Warrior 6 this is Battle Control. Repeating our six-niner-zero!" Came a frantic call. "Telemetry confirm an inbound, possibly armed with nucl.......SSSSQQQUUUEEEEEEEE."

"Holy shit!" Pageena yelled roughly pulling the headset off her aching ears.

"Aaaugh..." Yelled Marsha and Dafora in unison as their earpiece gave a similar squeal.

"What?" Jarob asked.

"Did I just hear what I thought I heard?" Marsha asked.

"I...I don't know?" Pageena replied.

"I don't either, but I'm glad I had witnesses." Dafora added.

"What...what was it?" Campbell asked.

"It...it sounded like Battle Control being attacked again another six-niner-zero. Then toward the end of transmission they said something about an inbound." Marsha replied.

"Yeah...they said it might be armed with a nuclear warhead. That high pitched screech we heard, may have been their transmitters melting." Dafora added.

"A what?" Jarob said in disbelief.

"I know it sounds weird, but that's what we heard. Right Daf, Marsha?" Pageena said.

The sound of a distant rumble, accompanied by short quick jolts similar to a minor earthquake brought them all to immediate attention. The pod shook for about twelve to fifteen seconds, then all was quiet.

"Take an exterior?" Jarob ordered after the quake subsided.

"EMP's up by about nine percent. Radioactive particle emissions have increased by sixteen. Those two trace signatures would accompany a nuclear detonation approximately 190 k away." Marsha replied.

"That's about the distance from here to Battle Control central." Campbell replied.

"EM is subsiding, rad emissions are holding steady." Dafora said.

"It's a good thing we're shielded, or when the first EMP went by all our electronics would have crashed." Marsha said.

"But that don't tell us what's happening now." Jarob answered.

"Are we gonna be ok in here?" Benni asked.

"Yeah...I think so...we pick up more radiation in space than we do here, even with this. Just to be safe though I'll turn on the exterior shields they installed for solar radiation. It may not help." Marsha said.

"But it wouldn't hurt." Dafora replied.

"Just the same, we should stay inside for the rest of the day, take a reading every hour." Pageena added.

"I agree." Dafora confirmed.

"Any ideas?" Jarob asked

"About what?" Pageena asked.

"About what...oh ha ha that's funny. About what happened funny lady." Jarob replied.

"Hey, I'm the tactical on this trip ask them." Pageena said.

"Well?" He asked looking at Marsha and Dafora.

"As near as we can make out, an air burst approximately 10 kilos was detonated in the same general area as Battle control." Marsha replied.

"How do you know it was an air burst and the size?" Campbell asked.

"We don't really. It just seems logical. You know as well as I that no one but warriors are allowed on Mars II. So Battle Control and all the ref stations are constructed on grav-platforms. The platforms are usually stationed about one or two k off the surface. We're still reading data, but so far it shows that the blast effect had a downward path. It wouldn't have that or at least to the extent we're reading if the blast had detonated on solid ground." Marsha replied

"Also if you'll check the map of that area the Ka-Halun mountains are right in the middle of the blast area. If the detonation had occurred on the ground, after the initial blast they're would be a downdraft off those mountains creating a keyhole effect in the radiation pattern. But, since the pattern is out and down like a ripple in a puddle, then the blast had to have been an air burst. As for the size, it would have had to be at least that to do any damage at all." Dafora added.

"But, Battle Control is, was, whatever, over 20 k square with no weapons at all. Why attack it and why now?" Pageena asked.

"You got me." Marsha replied.

"Whoever did it was very effective in their strategic use a

weapon of that kind. In fact they would have to use it or something like it, if they expected to succeed." Campbell said.

"Do you think they had time to contact Warrior Central?" Benni asked.

"Maybe but, I wouldn't count on it. If I were them, the subspace transmitter would be my first target. Cut off outside communication. We know they were under attack before the big one hit probably to take out the extended com lines, once they did that, who they gonna call local, that could help? Us, right?" Campbell said.

"How many...were on Battle Control?" Jarob asked.

"Hell I don't know. This time of the morning maybe 1200 all the ref stations were probably still in place." Campbell answered.

"A better question would be what are we gonna do now?" Pageena asked.

"Marsha, let's use the sat-link and inform Warrior Command of the situation. Find out what we should do now." Jarob instructed.

"I already tried. No good, the bomb ionized the atmosphere around us for a while. I'll try again in a few hours.

"I'm picking up life signs, about twenty k away, moving this way fast." Dafora broke in.

"How many?" Jarob asked.

"Large force, looks like hundreds." Marsha replied.

"Hundreds, that's impossible." Campbell said.

"Maybe, but that what it looks like." Dafora confirmed.

"Seal up the pod, put all systems non essential on stand-by and keep tracking them." Campbell ordered.

"Pods are sealed, all NE systems on stand-by, tracking.

They're fifteen k away and closing something kind of strange about their signature though. Either they're driving something huge, or there are thousands of them." Marsha replied.

"Any individual signatures?" Campbell asked.

"No, must be tight formation reads kind of triangular. Twelve k." Marsha answered.

"I can't see anything." Jarob replied as he looked out the front view port.

"They should be in sight any time now. Nine k." Marsha said.

"I still...wait...I think I see something...it's just barely visible, just their in the middle of the plain. See it." Jarob asked Benni ask they both strained their eyes to see any signs.

"I see some black dots, but that's about all." Benni replied handing Marsha a pair of binoculars.

"Six k." Marsha counted. "Holy God! Look at that!"

"There...there I see them now." Benni said. "Jeez...have you ever seen?"

"Not to clearly yet, but getting better." Jarob added reacting to Marsha and grabbing a pair of binoculars.

"Three." Marsha again counted.

"O my god! Will you look at that!" Jarob yelled.

"Holy shit!" Benni replied.

Computer End Recording...

Eight legs and all is hell.

In awe Benni and Jarob watched, the long procession of strange creatures that approached.

"Look at this...come here...come here, look at this." Jarob yelled.

Within moments they were all crowded around the viewport only to be amazed at the strange sight approaching.

Six humanoids sat upon six large eight legged beasts, swaying to and fro as they lumbered closer and closer. The beasts resembled spiders, though they were not arachnid, or insect. They resembled a mother spider who had gathered her young close to protect them. Six at first then another six behind them on and on they stretched, almost as far as the eye could see. Three formed a cluster in a tight triangle, with each cluster gathered into a tight pentagonal formation. Considering the bulk of each spider, the movement of the entire army was smooth and effortless.

As the army came closer, Pageena looked at the others. All had fear in their eyes.

"Oh shit!" They all seemed to yell together as they ran to get into their battle gear. Loud bangs and thumps accompanied their frantic brief lunacy.

As they tore open their battle packs, took out almost every weapon they had, and began either loading them, or starting their power up cycles. Then they almost flew into their protective armor, pulling on the fasteners of their tight leggings, quickly jerking on battle plates, and trousers. Finally, dumping on their helmets as they came out of their quarters, they each had at least two weapons in hand.

"Recorders on." Campbell instructed.

"Recorders up and running, exterior mics and sensor pickups activated." Marsha replied.

In the front of the armada a young woman atop the largest beast, rose up and began to speak.

"Jabbru aton....Jabbru aton. Me kenic, sal do shanni varook. Deesi ba dookas, mookta aku da haldashookba. Zashook en kenni Zashook. Umpjado kahalic da shodo thalna boocar, joopa, geldutinkas, da aku varook parapton fadie. Oomlatwee ka dalis kenni aku tauumakus. Kenic shanna mookta thalna joopa, varook ka dalic laporadaperanki. Jabbru aton, kilo mynki doro." These unfamiliar words tumbled out of the leaders mouth.

"Got any idea what she's saying?" Jarob asked.

"O shit...I forgot to tie in the translator." Dafora cursed.

"Varook de shookba ka." The voice continued; "There will be no more, unless we make it." The voice again chimed as the translator kicked in.

"Many years we have lived here, now we are prepared to die. We leave you now, but in two cycles of the sun we will return. If you are here you will be taken prisoner like the others if you resist you will die." She concluded.

She then again sat astride her beast, and with a flick of its reins, turned away. They all turned away until only a dark spot appeared on the horizon and the long well beaten path was all that remained.

"Tell me we have a recording of that." Jarob said.

"I said the recorders were on." Dafora insisted.

"Playback with a translator tie in." Jarob said.

"Yes sir!" Dafora answered sarcastically.

"Warriors within....Warriors within. We come to give you warning. I bring the Dookas, the Mookta and many others. War will bring more war. Battle Control is no more, refs, outsiders, or warriors are no longer safe here. Time is right for them to know of us. We send

you, tell them our Mookta killed the refs, and others will die if we are
to succeed. Warriors within, you cannot hide. Others use this planet
for War. There will be no more war, unless we make it. Many years
we have lived here, now we are prepared to die. We will leave you
now, but in two cycles of the sun we will return. If you are here you
will be taken prisoner like the others if you resist you will die."

"Well what do you make of that?" Marsha asked.

"Sounds like we got new refs in this game." Campbell said.

"Any chance of making it outta here in two days?" Benni
asked.

"No way the drop ship is programmed on a time delay. It's not
available until our scheduled war is declared ended by Battle control.
Since Battle Control was destroyed and we don't have the codes, I
guess we're stuck here." Dafora replied.

"How long...I mean sooner or later they'll realize something's
wrong...Right?" Pageena asked.

"Could be three weeks or more." Dafora said calmly.

"Three fucking weeks?" Jarob yelled; "Who's the dip who
thought up that scenario?"

"That's in accordance with section 247, paragraph C, sub-
paragraph 308, of the Warrior Command Rules. It states that no
attempt at rescue of any kind is allowed on Mars II, unless or until
such time as the scheduled war has concluded and the interested party
or parties are sure of survivors. If survivors are found, they are to be
held at Battle Control until they can be processed through and
provisions made for their safe return to their home world. There again
with Battle Control gone, it looks like we're screwed." Marsha replied.

"Anyone...anyone this is...If you...please respond." A voice
crackled over the com line.

"This is Warrior 6...you're about seven degrees in the delta 2
range, can you boost your gain? You're breaking up." Marsha asked.

"Boosting gain...there, is that better?" Came the reply.

"Still some break, but we read. Who and where are you?" Marsha asked.

"Names Jason Danberry, I was ref 245 for Battle Control. I was hoping they hadn't killed you guys along with all the rest!" He replied.

"Not yet anyway." Marsha replied. "Where are you?"

"There are three of us. I can contact the others, but only in line of sight. I think their long range com line's are down. We still have maneuvering thrusters, but everything else is gone." He replied his voice trembling.

"That still doesn't tell me where you are."

"Oh...ah...sorry. I couldn't really tell you, all scopes are down. But, I see a pod stack in the Cobi Cori foothills."

"That's us, just set your heading for that location we'll see you in a few minutes."

"Ok see ya."

"I don't know. It sounded like some kid fresh out of the academy." Marsha said looking at the others.

"It may be but he said there are two others beside him. Maybe one of them is a tech or something." Campbell replied

"So, that makes us over trained nurse maids right?" Dafora sneered.

"That's not what I meant. It's just that maybe one of them may have the codes we need to send for the drop ship." Campbell said apologetically.

"That's kind of a long shot ain't it?" Jarob asked.

"Better than the shot we got now." Campbell replied.

"Are our scanners still working? I mean there's no interference?" Benni asked.

"Scanners check out ok, but only on land sweeps. Anything above ten k and we're blind. Why?" Marsha replied.

"Just curious, I mean I feel helpless enough without knowing that the scanner's taking a crap on us too." Benni answered.

"I'm picking up our visitors on scan now." Dafora added;

"Their about 6 k out."

"Missile!...I've got a missile track! It's targeting lead ship of the inbound." Marsha broke in.

"Confirmed...missile track has acquired and locked on. Oh shit there's another one! Can't we do something?" Dafora yelled.

"We don't have long range capabilities, you know that." Jarob replied.

"Warrior 6! Warrior 6! I see a vapor trails...are they from you?" The panic stricken young voice demanded.

"Negative Jason negative. We read a missile lock. Can you evade?" Marsha asked as calmly as she could.

"I'll try." The nervous voice replied.

"He's pitching right three degrees. No good, missile still tracking." Marsha said.

"Kid...Kid." Campbell yelled; "Cut your mains for three seconds. Then power up to max for restart."

"If I kill the mains I'll drop like a stone."

"That's right...but do it anyway."

"Mains off." The young man replied as he felt the ship begin to drop rapidly.

"Ok...now kid, kick them back on full, set heading at one-six-zero and nose up 12 degrees."

"I've got the other two ships on track, they'll be here in ten seconds. You guys better go out and meet 'em." Dafora yelled.

"Mains are on but I'm blocked, I can't evade, somebody help! Tell me what to d..." A large puff of white and black smoke, and the report of the explosion some seconds later were all that remained of Young Jason Danberry.

The other two ships had landed only moments before Jason met his fiery end. Campbell and Jarob when out to meet them.

"Yo the ship." Campbell yelled as the ships began to purge their exhaust vents.

"I don't think they heard you." Jarob commented.

"Yo the ship." Campbell repeated after the system purge began to fade.

"Haloo." Came a voice from the lead ship as the canopy opened.

"Hi." Another voice intoned as the canopy opened on the second ship.

"Any word on 245?" The man asked as he hopped from the ship in the rear.

"He's gone." Campbell replied bluntly.

"Goddamn missiles. We're not supposed to have to deal with that shit." The voice of an older man grumbled.

"Campbell tried to twist him out, but two missiles are hard to get away from." Jarob replied.

"Bastards fired a second one eh?" The elder man asked rhetorically.

"Let's get inside." Jarob replied helping the elder man out of his ship.

"I'm Dathus…that's Bart. Thanks for the beacon we'd be dead now if it weren't for you."

Dathus was an older man, in earth years about mid 70s. With a wide face, familiar smile, and eyes as black as onyx, that seemed to reflect light from a thousand different locations. Bart on the other hand was tall, lanky, and had at least six shades of yellow in his hair. His steel blue eyes penetrated the squint he had developed over years of hard light exposure. From that it was obvious he was a reactor tech, for only they had this common characteristic.

"What happened?" Dafora asked bluntly as they entered.

"I'm damned if I know." Bart replied.

"Well, all I know is. Last night I was ordered to fire up the Linux RL20's, this morning. That's those piece of shit ships we were in they're usually in storage on the lower decks, we haven't used 'em in years. The commander didn't want to risk using an OS, or a ref platform. We were to make a perimeter sweep and report back. Check out some weird signals, they had picked up. Magnetic glitch of some kind nothing unusual I didn't get any feeling from them that the flight would be dangerous, that's why I said Danberry could come along. Catch a little more flight time on something closer to a real ship. Still, I feel responsible." Dathus said.

"We had just come off the flight line, when the first missiles came in and took out the entire sub-space communications deck. Dathus ordered us to open up and make a flat out run away from Battle Control. Jason went west, I went south, and He went east. About 25 k away we met up again and tried to home in any missile tracking systems. Another missile was fired and we were able to call Battle Control on emergency backup. Warn 'em you know. After that everything went dead. You could feel, I mean you could feel the heat.

30 k away and you could feel the heat. Then the first blast wave hit, all my scopes went dead. I saw Dathus's ship drop almost 500 k in less than a second. "Bart interrupted.

"Yeah and I almost peed my pants too." Dathus added.

"On the ground you could see the shock wave rolling like some king of gigantic snake, killing everything in its path. Anyway, our secondary's kicked in and now we're flying along flat out, full 600 kph, trying to outrun the next blast wave. Jason told us to drop down behind a range of mountains and let the blast roll over us. It worked. Then we followed him almost deaf and blind until we spotted you." "Bart continued.

"Lucky you got here after our friends left." Jarob said.

"I guess that depends on what you call lucky." Dathus replied.

"Yeah...hey I remember seeing this huge dust cloud as we were coming in. What the hell was that?" Bart asked.

"That's probably where the missile that got Jason came from." Campbell replied.

"Why? Why did the kid have to take point." Dathus lamented.

"His ship was the only one with an outside com line. Both of ours went out after the first wave. He relayed the coordinates of this place to us until we picked up a visual. Look I'm sorry he's gone to, but we can't change the past. So what do we do now?" Bart said.

"You may have jumped from heat to flame. That cloud of dust was an armada larger than anything I've ever seen. They gave us two days to get off "their" planet. You guys wouldn't happen to have the Battle termination codes so we could get our drop ship to take us outta here would you?" Pageena asked wryly.

"Nope." Bart replied.

"Sorry." Dathus said.

"Then I guess you could say we're all screwed, big time." Benni said.

"Marsha you're being awfully quiet." Campbell said as he noticed how Marsha sat, staring straight ahead.

"Oh...ah...sorry. It's just...there's something familiar about the woman we saw. I've been trying to think of where I've seen her before." She replied.

"Earth maybe? Not on Triton for sure." Jarob helped.

"No...no not Earth...wait...wait I know, she gave me the package with my orders in it on station 16. Names....Nina...Nina something...Cas...Casey...Cappi...Capr...Capri...that's it Nina Capri. She was the purser on The Delta Queen." Marsha remembered.

"What the hell is a purser on the Delta doing here?" Jarob asked.

"Looked to me like, leading an armada that could tear hell out of just about any army in the galaxy." Pageena replied.

"But, she speaks English, I heard her. I wonder why she didn't a while ago." Marsha inquired.

"It's not that important for now. What we should really do is try to find a way off this rock before...bef....Holy shit! The Waadii, I almost forgot about the Coonwaadii. They'll be here day after tomorrow. You guy's didn't happen to call them and call off the battle before your systems went down did you?" Jarob asked looking at Dathus and Bart.

"Like I said, the first thing that went was the sub-space communications deck. You know that when the deck went, the first thing Warrior Command would do is fly into the biggest bureaucratic panic you've ever seen." Dathus said.

"Then they would try, mind you try, to contact the Coonwaadii. It's doubtful though that the Waadii would listen to them though. Hell, I doubt that they even have their transceivers on. Right about now they're in the Toom Jaciaar. The "eyes of blood" ceremonies. A ritual blood bath they take right before a battle." Campbell added.

"You mean they take a bath in blood?" Dafora asked.

"It's symbolic of the blood of their enemies. Blood is God to the Waadii's. They bathe in it, drink it, and even use it to temper their metals." Campbell answered.

"I guess the pseudonym Blood Beasts is accurate then huh?" Marsha replied.

"Yeah, the briefing manual said something about Blood Beasts. But, I didn't think it was real." Benni added.

"Oh, it's real alright. Six months ago the Waadii were here, battling the Voorms. The war lasted all of nine minutes. When it was over the Voorms were torn to shreds and the Waadii were smearing their blood all over themselves. Yeck...disgusting." Bart said.

"Yeah...he...he they're still finding Voorm pieces over in sector 78. Bart there was the first one on the scene, I didn't think he'd eat again for a week." Dathus added.

"I almost didn't." Bart replied.

"Alright, I think that's enough about the Waadii's peculiar, hygiene habits. Let's get to a more urgent topic, shall we?" Pageena interrupted.

"Like what?" Benni asked.

"Like how the hell are we gonna get off this rock?" Dafora said.

"In case you forgot our friends on the giant spiders are due to come back here at about the same time the Waadii arrive. So, either we get out of here, or get our collective asses kicked so bad they'll be sifting the dirt to find what's left of us. With the size of that army, I'd say both we and the Waadii's stand to be in the same situation." Marsha added.

"No…no, the Waadii's would all be dead now. Their stupid, stubborn, militaristic attitude would have gotten them all butchered three seconds after they saw the armada." Campbell said.

The conversation went back and forth for a little while before finally someone came up with an idea. In the mountains about 5 k from Battle control, the original battle station had been constructed. It had been constructed before the Warrior program when the team sizes were unlimited, seven or eight centuries ago. No one knew if it had any power, or whether or not anything worked. But, up until a few years ago a tethering cable attached to the old Battle Control supplied some of the power to the new one and assisted in some computer upgrade operations. When the new one came on line full, the tether was dropped and the opening to the old Battle Control was sealed. Dathus knew that the power cells of the old system were probably still active. But, the only way to know about the rest of the gear was to take a look. So the decision was made. The next morning, Dathus, Bart and two of the members of the team would go and investigate, hoping the radiation wouldn't fry them the minute they arrived.

Computer End Recording...

New flies, old ointment

After some bickering bitching and general arguing it was decided that Campbell and Marsha would go with Dathus, and Bart. Neither of them liked the decision, but both knew that they were best suited. Campbell was chosen strictly because of his brute strength. Marsha, because she was the only one rated on the old style STAT 90, sub-space consoles.

If the thing worked, it's more than likely she was the only one who would be able to call up the security sequence they needed to get off-world. Though where they would go once they got off world no one knew and there was some protest, advising them not to split up. But, it was either wait for them all to be taken prisoner, or take the chance.

At the same time it was decided to move the pod cluster to a spot some 70 k to the east. With the coordinates mapped and plotted, each group now knew what they had to do and where they were going. The damage to the two ships was minimal except for the electrical systems, and most of the necessary parts could be scavenged without damage from the pods emergency relay panels.

Two hours of hard work passed, then three. Finally they were able to get all systems of both ships up and running, with only a minimum of phase distortion. Campbell and Marsha suited up in full battle armor and each took a nygor, with six clips of twelve shots each, and their blasters. Anything more and the weight could be a problem. Bart and Dathus were each given a blaster from one of the others' battle packs, goodbyes were said and they loaded into the ships.

Shortly after the ships had gone into the horizon, all the necessary preparations were made to move the pods.

"Lift off in two minutes. System is powering up now." Pageena said as the others strapped in.

"You got a lock on the target grid?" Dafora asked.

"Readings coming in now....lock is secure." Pageena replied;

"Main engine sequence active...get set people...firing...now."

The pod cluster began to rise slowly, then more quickly as escape velocity increased. Then a sharp turn to the east and the mains canted back to drive the entire cluster forward.

"Mains are on line and target sequence is engaged. E.T.A. reads eight minutes." Dafora said.

"Let's do a sweep of that zone just to make sure we're not going to fall right into enemy hands." Jarob said.

"Last known enemy position is in the opposite direction. But, just to be safe that's a good idea." Pageena replied.

"Zone reads clear." Dafora said.

"Landing cycle engaged, six minutes fifty-two seconds." Pageena said.

"No missile tracks or lock on information." Dafora reported.

"What do we do when we get there?" Benni asked.

"Sit tight I guess." Jarob answered.

"We could lay out some sensor augmentation pylons try to extend our range some." Pageena suggested.

"How many do we have on board?" Dafora asked.

"Five I think...no wait, only four. We had to scavenge one for parts...still, if we place them right we could extend range by about ten k or more." Pageena answered.

"I've been reading the target zone. There's a box canyon about halfway across the zone. Looks kind of like it's sitting on a stone shelf, if we alter our course three degrees we could set down there." Dafora said.

"How high are the cliff walls?" Jarob asked.

"O high...looks to be eighty meters or so. The shelf looks about a third of the way up. About thirty meters off the canyon floor." She reported.

"Think the shelf will support us?" Pageena asked.

"The shelf comes out of the side of the cliff but its solid all the way to the bottom looks wide enough too." Dafora replied.

"Could an enemy climb the cliff from behind and attack from above?" Jarob asked.

"I don't think so...see the mountain around it takes a 90 from about thirty meters or so up. If we set up our shields overhead to double, there's no way they could fire at us from up there and do any damage." Dafora said.

"Tell that to Magna station, they thought they were invincible too." Benni said.

"Still they'd have to get up there first." Pageena replied; "I say we go with Daf's heading."

"Change heading, .0031 degrees on my mark...mark." Dafora said.

Two ships darted into the early evening, it wouldn't be long now. The glow of the radioactive cloud was beginning to cease, still it was necessary to bring shields to maximum when passing through. They also knew that they would have to use kiznati barrier shields while on the ground until they had made their way into the cave. While the shields were built into the flight suits of the pilots, Marsha and Campbell would have to use theirs independently.

"I've got a visual on the cliff face where the cave is supposed to be." Bart said.

"Where away?" Dathus asked.

"At our five o-clock about two k. See that cliff down there on the right?" Bart answered.

"I've got it. Looks like a stretch of sand about 200 meters away from it. Let's set down there." Dathus replied; "Thrusters on hover mode."

With their thruster in hover they blasted their way toward the long swath of sand and gently touched down.

"Before we get out let me take a reading." Marsha said sweeping her scanner right then left. "Hmm...looks like the rad count is around .65. Still way to high to go without the shields, even with the shields I'd say we should limit our exposure to no more than 15 to 20 minutes."

"Got it, you guys copy that?" Bart asked.

"Yeah...limit exposure to 15 to 20 minutes." Dathus repeated. "Opening hatch...now."

The hatches of both craft opened with a hiss as internal pressure equalized. In a few minutes they were standing in front of a large wall of stones that looked as though they had fallen from farther up the mountain side.

"This must be where the old control room is." Bart replied.

"Looks like the top of that peak up there, was blasted off to cover the entrance." Marsha replied pointing to a plateau that appeared man made.

"Well, let's get to diggin' it out." Campbell said picking up a huge boulder and tossing it aside like a pebble.

Digging was hard, and would take longer than they thought. So, to make it easier and perhaps a bit faster, they turned the ships with the blast thrusters facing the covered opening. With a few blasts they managed to clear most of the smaller debris. After that a few large stones were all that remained. One was so large it had to be pulled out with a line and grapple attached to the bottom of the first ship. Finally, they were in.

"Check the auxiliary lighting." Marsha asked as she walked

into the darkened cave, turning on her belt-light.

"Where's that." Bart asked.

"Check the wall to the left of the opening. There should be a box." She answered.

"I...I don't see anything." Bart replied.

"Here let me take a look." She said walking to the opening; "Nope it's not here must be in the primary at the back of the cave."

Marsha walked toward the rear of the cavern, found the primary power system and after a few tugs on the lever, the power came on.

"Look at this stuff...it ancient." Marsha said as she looked at the room full of dust-covered, antique equipment.

"You think any of it will work?" Campbell asked.

"Well, I'll have to check out a few things but...we'll see." She replied.

"I just got a call on secure line. They've made the move, given us the location, and wanted to know how we're doing?" Dathus said.

"Well if this stuff doesn't turn to dust, evaporate, or just plain explode. I'd say we've got a shot, but I'm not giving any odds." Marsha replied.

"Anything we can do to help?" Campbell asked.

"Yeah...I've managed to power up the system, but the only terminals I can find are dummies. Check the second level, see if we can get a layout on any of the screens, if we can there's a good chance, the terminal ties to security." She said.

"I've found an old code book." Bart yelled.

"No good, they changed all the codes when the new place went on line." Dathus replied.

"Up here." Campbell yelled. "I got a screen with a complex layout on display. May not tie into security, but at least it may tell us where it is."

"The terminal is dumb but the location indicator show, security to be back here." Marsha replied after taking a quick look at the display and then proceeding toward the rear of the upper level.

They made their way through a small doorway and down the corridor to a large door labeled SECURITY—CLEARANCE AA109-AA229. The large rusted door presented a bit of a problem, corrosion had sealed in securely. But a couple of shots from Campbell's blaster melted away rust, and cut the main bolt. Then a couple of heavy shoves and it opened.

"The equipment in here is newer." Marsha replied.

"Yeah...they upgraded the security system pretty regular." Dathus replied.

"This system's only a few years behind what they had on the platform." She replied.

"Can you do anything with it?" Campbell asked.

"I don't know...Let's see...input my clearance code...check the data buffers...There...There's the code we need...It's an RL, easy to break if we need to. OK...I'm in. Contact with the ship established. Overriding time lock, then tell it that we've won and are ready for a pick up...I have a confirmation...Give me the new coordinates."

"Here they are." Bart said handing her the map.

"OK baby...here we go...new coordinates...pick up time...nine hours...We're going home!" She yelled.

A cheer rang out as the confirmation came in.

"I'll go tell the others." Dathus yelled running down the

corridor.

In a couple of minutes Dathus returned and told of the howls and whoops he had received when he gave the others the news.

"I told them to keep their scanners open so the ship could pinpoint their location by its beacon." Dathus reported.

"Do you think its ok for us to travel at night? Cause I'm ready to get outta this place." Marsha asked.

"Probably more ok than it is in the daytime." Campbell replied.

"Let's go." Bart said.

A few seconds later the first THUD shook the cavern, then another and another. Then voices could be heard from outside.

"There's a whole herd of spider things out there with all kinds of different critters riding 'em!" Dathus reported rushing back from the entrance.

"Is there another way out of here?" Campbell asked.

"Just the way we came in." Marsha replied.

"Then how we gonna get out of here?" Bart asked.

Before anyone could answer, the sound of blasters and screams could be heard from the direction of the entrance. Silently they crept forward, and peeked out of the entrance just in time to see a Coonwaadii warrior strike the last blow on one of the creatures attacking the cave. A few moments later, one of the Waadii approached, looked at Marsha and said:

"Are you one of the warriors we were to fight?"

"Yes." Marsha answered bravely; "You must be Bosrum, or is it Sokeiva? You're here kinda early aren't you?"

"I am Bosram, the question you ask is one we might well ask

you! We are not the fools you take us for. We too have noticed the many odd transmissions, and strange goings on here. My people were notified by your Warrior Program, we were sent for reasons of diplomacy. To assist you, and seek the truth of the mysteries of Mars II." He replied.

"Thank you for your help. But...." Marsha said.

"Before you deny us, here is a vid-chip made by our ambassador and Karen Trubin. It is for you, take it back to your camp and share it with the others in your team. We will contact you tomorrow at high sun." Bosrum replied handing her the vid-chip and turning away.

"Thank you..." She said weakly as they quickly got in their mag-car and sped away.

The flight to get to the new base camp was fairly short, and easy. However, since the ledge on which the pods sat was too narrow for them to land on, it was necessary for them to land at the base of the canyon and climb up to the pod cluster. Dathus bitched and complained about how he was too old to be doing this kind of strenuous stuff.

"Hey that's no excuse. See Campbell here? He's over five hundred and you don't hear him complaining." Marsha replied.

"Yeah...yeah...yeah. Sure he's over five hundred, and my granny is really, my grandpa in drag." He replied.

Bart puffed and wheezed but said nothing, and Campbell only smiled as Marsha tried to convince Dathus that Campbell really was as old as she had said.

At the top of the ledge, where the pod cluster was, they were met by the others and escorted inside.

"Guess who we saw?" Marsha asked.

"Who?" They all seemed to ask at once.

"The Waadii." She answered.

"What? What the hell are the Waadii doing here this early?" Pageena asked.

"Apparently the same thing we are." She replied; "Bosrum gave me this vid-chip he said they're ambassador made it with Karen on Earth." She continued as she approached the vid screen and slid the vid-chip into the slot.

In a few seconds Karen appeared with a Coowaadii standing beside her.

"Warriors...I'm here with ambassador Ruthma of the Coowaadii Confederation. We have produced this vid-chip together, to let you know this is no trick. A few days ago the Coonwaadii flagship Hespa was destroyed, by a beam similar to the one reported at the time of the destruction of Magna and Station 9. Though the Coonwaadii are not totally convinced, we have reached tentative agreement on a truce. One term of that agreement is that they are to assist us in your investigation. You will give them all information you have gathered so far, and cooperated with them in any way you can.

Though it is unusual for the Coonwaadii to come to us for help, especially at this time. We understand as they do, that if they were to take any hostile action even if they knew where it came from, it may have jeopardized the outcome of their Mars II battle with us. As you know rule 27 states that no planet may be involved in more than one conflict at a time.

We have received a communication from Mars II telling us that we are to cease and desist all warring on the planet within three days...or else. The order comes from a group calling themselves Vashomtu, or Deliverers. Somehow they have managed to eliminate all the monitors from all the systems in the Mars II agreement. In response we and our allies are taking all of our battle cruisers out of mothballs, upgrading them and sending them to Mars II. Some that were not in need of upgrading or required only small changes are now en route they should be arriving within the next twenty-six to thirty hours.

You are to keep in constant contact with Battle Control in your area and file regular reports with the cruisers when they arrive. They will contact you on secure link ZET red—blue--015. If you wish to contact the Coonwaadii, their call name is Dulak, channel Delta z, scramble to level nine. Both your teams will be on quick call, in the event of the need for immediate evacuation.

If you cannot evacuate or are captured you are to use K16 as an option. Karen Trubin and Ambassador Ruthma. Office of Intergalactic Intelligence JFR-1005980-5621 #00199 red-code-red-red-blue, File #109 BZET Alpha 919----END TRANSMISSION.

Computer End Recording...

Eyes of War.

"Guess they haven't heard about Battle Control yet." Campbell said.

"Guess not." Jarob replied.

"The question is what do we do now?" Dafora asked.

"What do you mean, what do we do?" Marsha answered; "We wait."

"But the drop ship will be here in another seven hours. What do we do when it arrives?" Pageena asked.

"I suggest we put Dathus and Bart on it and send it back up. I mean they're not military personnel. B'sides, we could use them as a link to the incoming Battle Group." Benni said.

"We don't really need a link. But, it's true they should get off Mars II before the fireworks start." Marsha added.

"I ain't goin' nowhere." Dathus yelled.

"Me neither." Bart agreed.

"Look, right now you two are in violation of Mars II rules. The only people allowed to set foot on this planet are warriors." Jarob argued.

"He's right, and as Earth allies we would be disqualified if we should resolve this problem and do battle with the Coonwaadii." Pageena added.

"I think the rules are pretty much moot now. You guys are gonna need all the help you can get." Dathus answered.

"Our help is on the way. Maybe if you saw the size of the army we're facing like we did yesterday, you might change your mind." Campbell said.

"What...a few hundred ragged creatures riding spiders...Yeah sure." Bart replied.

"Try a few thousand, well armed, well organized and I might add well led." Pageena said.

"How do you know how may arms they got? Or how well organized for that matter?" Dathus asked.

"First of all they used ground to air missiles to knock out OS's, ref ships and Jason's ship. Something tells me, that's just the tip of the iceberg. When they fired on Battle Control, the missiles they used had to be small enough to get past the proximity detection net. Yet, sophisticated enough to target, track and compensate. Missiles like that take maintaining, that's how I know we ain't dealing with a bunch of dummies. The military strategy in almost all battle situations is cut communications, power, and means of egress. Then set siege, in this case a tactical nuke made short work of that part. That shows they have the arms, no matter how old. Because they took out the sub-space before it could kick in and auto relay the danger, before they nuked the place. That shows organization. No...this was planned, in fact it was planned, and probably rehearsed before it was carried out." Marsha said.

"You could take that a few steps further and say whoever they are, they have an intelligence network. They knew what to hit and when to hit it. They waited until the next battle group arrived, and we all know information on which battle group arrives at what time is classified. Look at the pattern first, the OS's were taken out, kind of attention getters. Then warriors were taken, ideal as bargaining chips if they need to negotiate later. I think they destroyed station 9 and Magna to show their reach it's probably much further than those two stations. Finally, when they were prepared they destroyed Battle Control as a show of force. Or just to say they mean business." Dafora added.

"So, we have to get inside their skin, and see what the next move is. I got a bad feeling about this." Benni replied.

"Well, I still don't see why we gotta leave. Seems to me you guys are still gonna need our help." Dathus replied.

"B'sides, we could fly recon for you." Bart yelled.

"Yeah, recon." Dathus repeated.

"Arguing won't solve any of our problems. Dathus, you and Bart will get on the drop ship when it gets here. No more arguing!" Campbell commanded; "Pageena, you think you could fly one of their scout ships?"

"Sure, instruments are almost the same as this bucket." She replied.

"Then I suggest we stop assuming, and see what information we can get on our own. Tomorrow, we set up a recon party with the Waadii. Follow the trail left by their army yesterday. Pageena can do air cover and keep those sensors on wide sweep we don't wanna lose you. Then we find out the truth about what's goin' on here once and for all. Ok!" Campbell continued.

A round of "Yeah's" followed, and the plan was adopted. Dathus and Bart were fuming but there was little they could do about it. They knew they wouldn't stand a chance against the warriors if it came down to a fight.

The next morning dawned hot, with the air still slightly tainted with residual radiation. Marsha's morning sensor sweep showed that the rad count had dropped dramatically in the night.

It was still high enough for them to require protective gear when leaving the pods. Shields were not necessary though, and it was found that their battle armor would more than protect them. The drop ship had arrive before anyone was awake and taken a position at the base of the canyon. Dathus and Bart were hustled inside and soon the ship roared off into the morning sky.

Though they tried several times, contact with the Waadii was not made until high sun. Just exactly when the Waadii had said contact would be made. Plans were made to rendezvous with them in a small glade, near the end of the plain at the base of the canyon. From there they would break up into three parties of three, leaving one party of two to guard the rear. Bosrum, Marsha and Jarob were team one. Sokevia, Dafora and Uberus made up team two. Campbell, Jimba Ti and Tinnic Ba were number three, Benni and a new member of the Coonwaadii team Uum-Daluente would guard the rear, while Pageena would fly cover.

Not much was known of Uum, except that he was Reluthian, and had been conscripted from one of the many worlds the Coonwaadii had annexed over the centuries. Each party would pace no further than 50 meters from the others, making sure to keep both visual and verbal communications open. Nineteen grueling kilometers later they came upon a large stone wall, over thirty meters high and stretching further than the eye could see in both directions.

"God! I wonder why no one ever reported this structure. There's no way anyone could miss it." Marsha gasped.

"Maybe they could. This looks like a recent structure to me. See the striations in the stone, it's been recently worked. At least this section couldn't be more than a couple of years old if that." Jarob replied.

"Team two to team one do you read?" The com system cracked.

"Team one this is team two, is your channel secured?" Marsha replied.

"Affirmative, we are secure. Suggest we run a secure conference line." Dafora replied.

"All lines open and scrambled com to com. Are we clear with everyone? Com check...one...two...three...zebra...alpha...boy?" Marsha asked.

"Team two copy and on line." Dafora answered.

"Team three, link on line." Campbell replied.

"Back up is on line." Benni answered.

"My readings indicate a sensor array that runs the entire length of the top of this wall, strange though they don't register the wall at all. My guess is they know we're here." Marsha said.

"My sensors indicate the same thing, no real substance. They're doing massive sweeps every fifteen seconds or so. Doesn't feel like stone though. Anybody see a door?" Dafora replied, rubbing the stone wall, search for a switch.

"Just wall." Jarob answered.

"Anyone have a suggestion about what we should do now?" Campbell asked.

"COME INSIDE!" A loud voice demanded as the wall began to undulate, then separate and form a large arched opening.

"Team two, join us at our position. Team three and backup remain on station. If we don't report back to you in one hour, repeat one hour, you are to go back to the cluster, call the drop ship, leave and report our findings to Warrior Central. Copy that?" Marsha ordered.

"We copy, remaining on station one hour, return to cluster, call drop ship, depart and report all findings to Central...is my readback correct?" Benni asked.

"Correct team one and two out." Marsha said as team two arrived; "Team one to Condor, can I have a flyby at transmission coordinates?"

"Coming up on you now." Pageena replied.

"Tell the pilot to back off, or we will destroy that ship!" The voice again commanded.

"Condor, this is Team one, suggest you join team three and backup, and remain on station." Marsha relayed.

"Agreed, I picked up the message here too. Take care. Condor out." She replied turning the ship sharply.

With some tension they made their way through the archway and stood in awe at the strange sight before them. The fortress was immense, well over 100 kilometers square. If they could have seen it from the air, they would have seen a five pointed star pattern, with three repetitions of that pattern, spaced about fifty meters apart toward a central complex. Once they were inside the wall sealed behind them as though no opening had appeared at all.

"You were given two cycles to leave this place. Yet you come here, Why?" The voice demanded.

"Well you were so kind when you visited us we thought to return the favor." Jarob replied bluntly.

There was no answer to his flip remark, unless you would call the immediate insurgence of hundreds of angry looking soldiers perched on their arachnid steeds, an answer.

"Who is your leader?" The voice asked.

"We have no formal leader...But I will speak for the others if they agree." Marsha replied.

"I do not agree...I am Sokeiva Pocto of the Coonwaadii Confederation. My brother and I have come to talk peace, but we are willing to back that up with War if necessary." He replied moving to Marsha's side.

"Your arrogance does justice to the reason you are called Blood Beasts. The one beside you, her name is?"

"I am Marsha Tate of the Office of Intergalactic Intelligence. Who are you?" She asked.

"And the other of your kind, the one with the quick tongue?"

"I' Jarob...Jarob Ka...from Earth."

"Your friends will be given food, water, and a place to rest. You three follow the escort now approaching, if you stray beyond the path they set, you may be injured."

"Marsha, you think you'll be ok?" Dafora whispered.

"I don't know. But, I think we'd better do as they say." She replied.

Sokieva on the other hand, had a bit more of a physical response to their demand. As he drew is blaster to fire, three shots came from three different directions. The first tore through his left arm above the elbow, severing it. The second struck him in the left thigh, knocking his legs from beneath him. The third struck his brow and Sokieva, was an ex Coonwaadii.

"Stupid foolish creatures, we allowed you to keep your weapons thinking the size of our forces would reflect the futility of such rash actions. Now we must take them from you, place all weapons on the ground in front of you." The voice yelled as the army pressed in aiming all weapons at the remaining group.

Bosrum was just about to repeat the actions of his brother, when Marsha touched him gently on the arm and suggested he do as they say, unless he wanted to get them all killed. Soon all their weapons lay in a stack on the ground.

"Now follow the escort as instructed." The voice commanded.

Bosrum insisted he join Marsha and Jarob in his brother's place. Together, they followed the escort as the wall ahead of them began to move and form an arch similar to the one used before. As they looked around, the size of the city's population overwhelmed them. The streets were clean and well kept, shops and stores were well cared for. Children played happily in the streets, and an air of contentment was all around.

As they passed through the third wall, before them stood several large buildings stretching almost half a kilometer into the sky. Oddly they could see the buildings from beyond the second wall. It was thought that they may be communications towers. However, upon closer inspection, the buildings were central trade stations, distribution, and social coordination centers.

With the guards leading the way they entered a large building that took up almost all of the lower part of the stars five o'clock position. A large steel door groaned open they entered and stood in a huge room with high stained glass windows, large marble columns, and highly polished floors. A small desk sat in the middle of the room, four chairs encircled it.

"Sit here and wait!" A guard commanded pointing to the chairs.

Marsha, Jarob and Bosrum sat in silence, waiting. Only a few moments passed, an elderly man with long silver hair and bright blue eyes entered, accompanied by the same young woman Marsha had seen both on the spider creature and on station 16.

"Wh..." Marsha tried to ask.

"Ah...ah..." The man replied holding up a bony finger.

He then took a seat behind the desk and the young woman took up a standing position behind him to the right.

"I do not think you will understand what I am about to say to you. But, I will try it anyway." He said.

"Who..." Marsha tried again.

"Ah...ah...do not speak. Your voice will not be heard here for now." He requested.

The young woman then leaned down and whispered something in the old man's ear, he whispered something back and she left the room.

"Now, first I must tell you that we are here, we have always been here, and always shall be. Next, if any attempt is made to set up further wars, we will not allow it. We control all the monitors, and have enhanced their range and destructive capabilities far beyond their original specifications. As example of their force I give you Magna and Station 9. Using Nalvoron transporter technology, we are able to instantly place a monitor in any sector, at any time, for any purpose, under our complete control. If you will look at the vid-screen behind me, you will see the monitors in clusters of ten awaiting transport. Their targets are each of the Planets within the Mars II alliance, all of their training facilities and all ships in each battle fleet." He said with a smile.

"Ma...may I speak now?" Marsha asked.

"Yes." He replied.

"Why. Why are you doing this?" She asked.

"Because Mars II is my home, it is our home. The home of the Mookta to the south, the Dookas to the west, Aperitagleon in the north. This is our capital city Savor-laropis the City of Rest. We have been here always, behind our holo-walls, though they have not always looked as they do now. Before my time...before my father's time...and his time before that, our descendents left behind after many a great war when war was more than the game you make of it now...we have been here."

"When those wars ended, no care was taken to see that all who had fallen were dead. Those left alive, were left behind. Living in the radiation among the torn bodies of many a violent dead they scavenged the technology at hand. Existing on the foul remains of the unfortunate ones who had fallen in battle, fighting the vultures for one more mouthful to sustain themselves yet another day. Those who survived built what we have now. Those who continue here, continue to build. This planet is no longer what it was made to be. Through our generations, we have earned the right to live here in peace. We will live in peace, whether that peace is imposed on others or agreed to willfully." He answered.

"Who are you?" Bosrum asked tersely.

"Na-Tupoc, is my name. I am High Elder of the city. Leader of the Ninth Council, and for all intents and purposes the only one with whom you may speak."

"But surly you must realize the tremendous value of Mars II, both in terms of its construction, and operations?" Jarob replied.

"We know only that it is our home, and we will not let it be used for war anymore." Na replied.

"You cannot defeat the forces of the entire Galactic Alliance, it is we who have made Mars II, so that we could settle our disputes off our own worlds." Bosrum replied.

"Why?" Na asked.

"What do you mean. Why?" Bosrum yelled.

"I ask you why? You have built this place of beauty, and serenity. In the times when you are not destroying it, it is more beautiful that you could possibly imagine. Yet, blinded as you are by your lust for war, you cannot see what is there before you." He replied.

"I have seen the beauty of this place. I have seen the Jopro, and the moons rising in evening time, behind the mountains and seen the tall grasses wave in the breeze of this summer's day. I know the beauty of this place. But, I have seen similar beauty on my home world. Would you have us destroy that when we have built this place to save our world from that horror?" Marsha replied.

"War is not a pleasant thing. It is not meant to be a pleasant thing. What you have made it, is a game. A game in which the act of war is PLAYED...PLAYED! As children in the fields would pretend that death could not hurt them. You send your teams, play at war for a while, and after a winner has been chosen you go back to your secure lives, secure homes, and forget that the horror of war is what makes it a thing to be avoided." Na said.

"We all understand that war is inevitable. There are times when it simply cannot be avoided." Jarob replied.

"Nonsense! War can always be avoided." Na yelled.

"What about you? Don't preach to us about war and avoidance. What you are doing now, the violence you have committed, the war you have declared. That's not an attempt at avoidance." Marsha replied.

"True, we give you back your horror. To show you that war is not a game, not a fantasy that makes settling your arguments easy, quick, and painless. You want wars, do them on your own worlds, or make another, but do not make the same mistakes again. There will be no more wars here!" Na responded.

"I will not tell the others in the Coonwaadii Confederation, that we can make no more war here!" Bosrum yelled.

Within seconds Bosrum had jumped to his feet, a small concealed blaster in his hand. Marsha tried to grab his arm as he took aim and fired. Na-Tupoc, was struck in the chest by the discharge as he to rose to his feet, then fell across Jarob's lap. Seconds later, the young woman burst into the room. Bosrum turned, and aimed the blaster at her as she entered. He fired once, then again, missing both times. She returned fire as he ran to a large marble column near the windows at the back of the room, wounding him slightly in the left calf. Marsha went to Jarob and the old man in an attempt to help shortly after he fell. When the young woman came in, Jarob pushed the old man off of him and motioned for Marsha to join them on the floor behind the desk.

As Marsha and Jarob peeked over the desk, guards rushed in from everywhere. Bosrum fired again and again, taking position behind the column, fleeing his cover only briefly to continue his assault. Three guards went down as he fired over and over, sweeping the room with one blast after another.

"Set weapons on full kill!" The young woman yelled as she took cover behind a wooden panel just to the left of the door.

Though now he had been struck by many a glancing blaster shot, Bosrum would not yield. Outnumbered, and surrounded, he continued his siege, and when his blaster was nearly spent, he pulled a large dagger from another compartment of concealment and charged at them full force.

"Now I will have vengeance. VENGEANCE!!!" He screamed

as loudly as he could, running toward his foe, firing and slashing.

In seconds he was struck with many shots simultaneously. As beam after beam impaled him, he melted away until all that remained were the echoes of his screams and the final report of his blaster, in the cavernous room.

Marsha stood, and turned her attention to Na, his head laying in Jarobs lap. Jarob faned him with his hand and gave him an occasional light smack on the cheek, trying to wake him.

"He took quite a jolt." Jarob said looking up at Marsha.

"Slapping him like that won't do him any good either. Jeez where did you learn first aid?" Marsha replied.

"Hey, I ain't no med-tech. I puke when I see my own blood." Jarob replied.

"Oh...there's a pretty picture." Marsha quipped; "Here, let me see. Doesn't look like any real damage, just blaster burns and shock wave concussion."

"Lucky the blaster charge is laser light, looks like most of it was deflected off this light colored, heavy, robe." Jarob replied as Na began to come around.

"How 'bout them. They alright?" Marsha asked one of the guards as he helped others to their feet.

"Yeah...Ok...just a little shaken, and a few minor burns, nothing major." He replied.

"Father..." The young woman replied running to Na's side.

Computer Recording Off...Buffer filled...

BEGIN TRANSMISSION...

TRANSMISSION COMPLETE...

War is war and hell is hell.

"Now you see that we are not the fools you may think us to be." The young woman replied, helping her father to his feet, turning her blaster toward Jarob and Marsha.

"Wait!" Na yelled shaking his head slightly, giving Jarob a thankful smile.

"Why? I told you we would not be able to control these savage creatures." She replied.

"To do that would make us no less the savage. Do you want to be the thing you hate?" Na asked.

"No...I am sorry. Master Na." She replied.

"I...I know you." Marsha replied looking at the young woman. "You're Nina...Nina Capri. The purser on the Delta Queen, wha...what are you doing here?" Marsha asked.

"You might say we have the same task. Yours to find information, mine is to take it." Nina replied.

"The vid-chip, you are the one who gave it to me." Marsha said.

"I am also the one who copied it." Nina added.

"That's impossible you can't copy a secured vid-chip. The crypt encoder would destroy it before that happened. Even if you did manage to copy it, the security scan encoded into it would kill anyone who tried to use it." Marsha answered.

"Not if you have the encoder key which I do. Honestly, the warrior security codes are a joke. My kid sister could crack them with a second grade reader." Nina said.

"How did you get our coding sequencer and briefing numbers?" Marsha asked.

"Didn't need them, you guys follow the line too straight. All I had to know was the first opening output. From there knowing the sequence and the color, number, or drop code was just a matter of guess. I took your whole security net in about ten minutes...sloppy...just plain sloppy." Nina replied.

"So that's how you knew when the next battle would be fought, and in what sector." Marsha replied.

"That is unimportant now." Na said.

"Yes...yes you're right. I will proceed with our third step." Nina replied leaving the room.

"What's the third step." Jarob asked.

"Come, we will gather your companions, then all see the third step together." Na replied; "Call the ones who did not come into the city and tell them to meet us at the main gate. Have no fear no harm will come to any of you unless you bring it upon yourselves as Sokieva and Bosrum did. We will send an escort to ensure their safe arrival."

Marsha contacted Pageena, Campbell, and Benni explained the situation and told them to expect an escort any moment now. At first she suggested that the Coonwaadii warriors stay behind to avoid any further problems. Uberus, Jimba-Ti and Tinnic-Ba, were less than happy with that arrangement and insisted they be included. Then they were told by the escort when they arrived, that staying behind was not an option.

"Scan them thoroughly, if you find any hidden weapons take them, if they put up any resistance. Kill them!" She heard a guard reply.

There was no resistance, and soon the warriors were astride the large arachnid creatures they had grown accustomed to. The smell however, was a different tale. In but a few minutes the entire team was united and walking with Na down a long wide corridor in the main building, in the center of town. Tinnic-Ba complained in the pig grunts and muffled squeals, his race called language. But, it did him little good. In fact the universal translators built into most of their battle gear, was unable to decipher most of what he meant. The odd "fuck" or the high pitched "asshole" were about all that could be made out, the rest was static so most of them turned their translators off his frequency.

Uberus and Jimba-Ti were angry and looking for revenge until they saw the puddle that was once Bosrum. Revenge didn't appear that important to them after that. Uum-Daluente, was quiet, but then he (or was it she) was always quiet. Tinnic-Ba finally shut his mouth and walked in silence with the rest of them.

Four security stations were within the corridor, spaced about thirty-five meters apart. Each one sealed by a large door of Trianium and Novalite.

"I've never seen metal made of these two compounds before." Marsha said.

"There are many things you have not seen." Na replied coyly.

On they went past the third and forth security doors, into a large complex of rooms all within a central area. Once in the core area, the entire complex began a long descent, resting on a massive stone base some fifteen kilometers beneath the Mars II crust.

"When Mars II was found many ages ago it was believed no life existed here. In truth however, ancient wars so long ago the mind cannot conceive reduced this planet to what was found upon its discovery. These wars forced the inhabitants underground, into areas such as these. Being a xenophobic race, they had no means of escape, or inclination to do so. Rapidly, the biosphere failed, and they found they could not reproduce the atmosphere quickly enough to regenerate it. They died; all of them. The technology you see here is all that remains of them. Though vast in its ability, it is easier to use than a child's toy, with more destructive capability than thousands of anti-matter explosions.

With that technology we were able to create the holo-wall.

Solid, but not solid, and capable of being manipulated into any shape we desire." Na explained as they all looked in awe of the gigantic machines so ancient, yet still in operation.

"Why didn't you give this information to the Council?" Marsha asked.

"Oh there's an idea!" Jarob smirked.

"What? What's that suppose to mean?" Pageena asked.

"I mean what do you think the Galactic Council would do if they knew this type of technology existed?" Jarob asked.

"Open it up to further investigation, and research." Dafora replied.

"You wish!" Campbell replied.

"But, they already know!" Na interrupted.

"What? They already know about this place?" Marsha asked.

"Yes. They also know about us. They have known for many years." Na replied.

"If they already know all about you, then why the hell are we here?" Pageena yelled.

"You don't understand do you? We stand between the Galactic council and the technology you see before you. I would guess you were sent here to assess the size of our forces and tell them where they may strike for optimum effect." Na said.

"No we were sent to fight the Coonwaadii." Dafora said.

"There is no threat from the Coonwaadii." Jimba-Ti said.

"But, what about the disputed quadrants?" Pageena asked.

"Don't you know propaganda when you hear it?" Jimba-Ti

replied. "There has been no action against the Earth Alliance by any of the Coonwaadii Confederation. We too have heard the stories and wondered why in fact war had been scheduled between Earth and the Coonwaadii, when none had been petitioned."

"Then why did you come here?" Campbell asked.

"For the same reason you did. To find out once and for all what the hell is going on." Jimba-Ti replied.

"Ok Na, what the hell is going on?" Jarob asked.

"I assure you I don't know. I have on many occasions sent vid-chip messages to Karen Turbin the Chair of the Council and have gotten few replies." He answered.

"You...you've contacted Karen?" Marsha asked amazed to be hearing what she was hearing.

"Yes many times." Na replied.

"Why?" Marsha asked.

"I met her at a diplomatic function a few years ago, we spoke and she seemed genuinely sympathetic to our cause." Na replied.

"Years!...Karen knew about this years ago. Wait a minute...wait a minute, let me get this straight. Karen has known about you and your movement to stop wars on Mars II for years?" Marsha replied.

"Yes...and I have sent her a message of intent before each action was taken." Na answered.

"You mean she knew about all the kidnappings, all the station destructions?" Marsha asked.

"Yes and three days before we destroyed Battle Control, I sent her another message." Na confessed.

Marsha found a chair and sat down before she fell down.

Campbell, Dafora, Benni, Pageena, and Jarob, were all equally dismayed at this newfound information.

"I thought you knew these things!" Na replied not understanding their reactions.

"No...no we didn't." Campbell replied.

"You mean this Karen didn't bother to tell anyone of our position on Mars II?" Na asked.

"Not a word." Jarob answered.

"Then those people...the people on Magna and Station 9...they all died...with no warning from the Council?" Nina asked.

"Yep...and battle control too if they had been warned in advance. Do you really think you could have destroyed them so easily?" Dafora replied.

"Thinking back...it did all happen a bit easier than we had planned." said Na.

"Co'mon...I don't wanna put down your strategies, techniques, or timing. But..." Campbell said.

"There is only one weapon we used that has no counter." Nina interrupted

"What's that?" Marsha asked.

"The monitors we enhanced with technology we recently found here." Nina replied.

'When did that happen?" Marsha asked?

"A few months ago it's what we used to attack the stations. Some new tech stuff we got from these computers. It fires a sub-space sensor beam with an active nucleus that can be deflected off any surface exploding on contact. Because it's a sensor it can monitor itself in flight and increase the charge to whatever strength is needed to destroy its target. Nothing can stop it, once it's fired and nothing can withstand it when it makes contact." Nina replied.

"You call it an enhanced monitor?" Pageena asked.

"That's how we conceal it. It's small enough to fit inside a monitor. Using Nalvoron transporter technology we have already positioned almost all of the monitors. The last are on their way now." Na replied.

"What? I thought you were going to place them, you didn't say they were in place already?" Jarob said.

"We lied!" Nina replied.

"Are they active?" Marsha asked rushing to a computer console to get a bearing and readout.

"Not yet." Na replied.

"Before you activate them let me try to contact the council. Ambassador Quant, of Amna, should still be there. If I let him know what's going on we may be able to avert any more killing." Pageena said.

"Ok we'll wait, but not long." Na replied.

"Oh shit...Oh shit...Oh my God...!!!" Marsha yelled; "You didn't study the spec's on this enhancement very carefully did you?"

"We studied them. There were a few things we didn't quite understand. But, we studied them." Na defended.

"Do you know what interactive decay is? Cascade homing system? Anti-matter vent discharge? Convergent targeting to source?" Marsha asked.

"Holy shit!" Dafora repeated as she sat at the console beside Marsha.

"Interactive decay is the decay of the nucleus over a certain distance. I don't know what the other things mean. What are you getting so excited about." Nina replied looking at both Marsha and Dafora as if they were crazy.

Eventually, Pageena, Jarob, Campbell and Benni gathered around the console. The remainder of the Coonwaadii team did not share their interest. So, much of their time was spent looking at the main weapons console, not the monitor console.

"You're right about the interactive decay, but the decay on this is .025 per 30 light seconds. That means for every 30 light seconds the active nucleus will lose, .025 of its mass either as spent fuel plasma or as overflow discharge through the vents. But, the mass has to be replaced or the monitors fizzle out and do nothing. Part of its active matrix, is to replace lost or spent fuel with hydrogen, through the re-animantine net. That's where the change was made. Hydrogen is replaced with anti-matter. Micro interactions between the anti-matter and matter are what regenerates its fuel supply. increasing its range and strength geometrically as it moves throughout the galaxy. Like it normally would, the machine would route any residual particles captured in flight into its discharge vents to be released on impact. We're talking a matter/anti-matter explosion." Marsha explained.

"Exactly, total destruction. We must make sure no one forgets this is our world now." Na said.

"I don't understand." Campbell said.

"It's simple really, look the sensor read out gets it's reflected input, reads any increase and bumps the hydrogen bussard collectors to increase the amount of compressed hydrogen in the charge head. Then on impact the two sub-critical masses in the reaction chamber collide and boom, you have in essence a hydrogen bomb. The greater the hydrogen mass it manages to store, the greater the kaboom when it finds its target, nothing could withstand that, it would be like having a sun blow up in your face. In this case however, the reaction chamber has been altered becoming an anti-matter containment field. The weight has also been altered, because there's no need for the sub-masses to create an explosion. That happens when anti-matter and matter mix on impact." Dafora explained.

"Dafora, can you call up a grid display of the entire area covered by monitors?" Marsha asked.

"I'm way ahead of you, that's coming up now." Dafora replied.

"What the hell are you two so excited about?" Campbell asked.

"We've got some deep serious shit coming down." Marsha replied.

"Like what?" Jarob asked.

"Like these assholes are about to punch a hole the size of God in the universe we know and love." Dafora replied.

"Well put." Marsha complemented.

"What's that mean?" Pageena asked.

"Give me a view of the monitor locations Daf." Marsha asked; "It means that these idiots just set 1,067 matter/anti-matter charges all on a homing course and their coming right back here." Marsha replied.

"Here? What are you talking about, they're targeted for the locations shown on the vid-screen?" Nina shouted.

"The transporters were set to those coordinates! All the monitors are set on a homing frequency to return to point of origin, that means right back here! You fool! That's what cascade homing means." Dafora added.

"How far away is the closest one?" Marsha asked.

"Sixty-four light years estimated impact eighty-nine hours." Dafora replied.

"No it won't impact when it arrives, convergent targeting requires that all monitors stay in station keeping until the remainder are on location. Once that happens they coordinate target information then..." Marsha added.

"It's bad enough to have one matter/anti-matter explosion. But, 1,067 such explosions could very well destroy every system in this spiral arm of the galaxy." Campbell said.

"The explosion is only the final step. At the rate of anti-matter build up these monitors will have. The chances are they will all vent large quantities of anti-matter into the various sectors of space, making nice little booby-traps for anyone or thing that happens to be in the area or haplessly passing by." Dafora explained.

"What can we do?" Pageena asked.

"I don't know...I don't even know if we can do anything." Marsha replied.

"We're lucky in one regard." Dafora said.

"What's that?" Jarob asked.

"At least they're moving away from most of the trade routes and into a less densely populated region." She replied.

"It looks like we may lose about fifty of them in the Paladian asteroid belt. This group here should contact it in about ninety hours or so." Marsha said directing their attention to a rapidly converging cluster of monitors moving toward a large band of dots on the scanner screen.

"We better contact everyone we can this thing has gone far beyond Mars II now." Jarob suggested.

"I agree." Benni confirmed.

"Where's your sub-space communicator?" Benni asked Na and Nina tersely.

"It...it's over here." Na replied walking over to another console to the left of the monitor scanning system.

"Wait—let me call Karen!" Marsha insisted, overturning her chair as she swiftly moved toward them.

In a few second the console was activated, since there were no monitors around Mars II to use as sat links. It was necessary for them to relay information through the onboard sub-space relay of the warships now in orbit. After getting through all the coded secure bull, they were finally connected with Warrior Command and placed in a holding cycle until Karen was available.

"Ambassador Turbin here....Marsh....it's nice to see yo..."

"Don't hand me that shit, you bitch!" Marsha yelled.

"Marsh...hon..I." She tried to say.

"Don't bother! I know Karen...we all know and soon so will everybody." Marsha yelled.

"Know?...know what?" Karen replied innocently.

"Don't put on the innocent face for me! I just found out that you knew all along, about what was happening here on Mars II. That you knew about the attack plans in advance, and did nothing. How many Karen? How many beings died because of you? Thousands maybe tens of thousands?" Marsha continues as her anger welled.

"I was under orders!" Karen explained.

"Orders? What kind of orders justify the death of thousands?" Marsha yelled.

"I can't tell you." Karen calmly replied.

"Oh come on! Is it because you can't get your hands on this technology? Or is it that the higher ups want to take it all for Earth and leave the other planets in Alliance out?" Marsha replied.

"I...I'm sorry I can't..." Karen replied.

"That's it isn't it...the Earth Council wanted all this for themselves...isn't that right?" Marsha insisted.

"As I said I can't respond to that." Karen replied; "Wait, I've got another call, stand by."

"No...Karen don't you dare...!" Marsha yelled as the screen switched to stand-by; "Damn!!"

A few moments later a new face appeared.

"I am President Daryl Lemming of the Earth Defense Directorate. You and your team are to remain on station until such time as all data and equipment from the Mars II artifacts can be removed and shipped to our main Research facility here on Earth. Karen Turbin has been relieved of all duties pending an investigation into her part in the cover up of the findings on Mars II, and the subsequent deaths related to that cover up. Rest assured that all possible action will be taken. You must also understand that neither I nor the Planetary Council had any idea what was happening. Of course we knew of the artifacts found there and steps were being taken to investigate further. Karen Turbin and her followers are solely responsible for the inaction of this department when all attacks took place. We knew from some intercepted communiqués that Mister Na and his soldiers had kidnapped a few warriors, this was interpreted as a ploy for negotiations, and we were willing to talk. Is Mister Na there?"

"Yes...I am here." Na replied.

"Sir...It is my pleasure to meet you. You may not understand fully what I am about to say, but you and your compatriots are to be moved from Mars II. A new planet has been chosen for you and it is yours to colonize as you see fit. Unfortunately, in light of the attention and heavy scrutiny Mars II will undergo from now on. I fear it will not be a place for you and your people to live in peace anyway. All wars on Mars II are canceled until further notice. With the sudden disappearance of the Monitors I must also tell you that wars have broken out on six planets in the last 20 hours...."

"I hate to interrupt you. But, we know where the monitors are." Marsha broke in.

"Where?" The President asked.

"Right now, they're all headed back to Mars II, and I would say that about four hours after they all get here, Mars II, your artifacts, and anything alive for about forty-thousand parsecs or so won't be around anymore." Dafora replied.

"Why's that?" He asked.

"Because these morons enhanced them with technology they gleaned from these computers and teleported them to all the Warrior facilities, all Planets in the Warrior Alliance, and their related training camps, in a power play to stop wars on Mars II. They didn't realize that once evidence of a new technological species was found here it would automatically halt all wars until it could be investigated, according to the 2178 archaeological charter." Pageena piped in.

"I'm afraid I still don't understand?" He replied.

"These people tried to use some of this technology, to create a control series of monitors that were vastly more powerful than the normal monitors. I'm sure they thought that if they could control all the monitors with this new weapons upgrade they would be able to live in peace. But, we all know that just the administration of such a system would be anything but peaceful." Marsha explained.

"So mister Na! You and your friends just couldn't stay away from the central chamber could you. You just couldn't be content with your holo-walls, molecular replicators and armament facility. You had to go into the central chamber, even though you were warned not to enter areas, or deal with things you may not be able to understand. Our engineers would have been there in a few days, why couldn't you wait?" President Lemming demanded.

"You have no right to lecture me as you would a child. We have been waiting patiently for years while you, and your diplomatic corps, have avoided, and otherwise dismissed our sovereign right to exist in peace here. Now you offer us a new planet to placate us, until when? Until you find something of value on it, then we shall once again be asked in the interest of peace to leave. Thanks, but no thanks. All of the actions we have taken thus far were duly noted, and the petitions of intent were filed with Ms. Turbin.

What she did with those petitions and whether or not you were notified is, and was, not our concern. We demand action now! We have been patient long enough. As for what we have done, it is regrettable, but nonetheless we are willing to destroy this world rather than see it used for more death. Can you claim the same for your world? There is an old saying on your world. "War is hell." We disagree war is war, and for too long you have made this world an arena for your bloody games. Yes Games! War is not meant to be a game, it is meant to be avoided. You have made it a thing of convenience. No one on your home worlds has died in a war in centuries, except those you send here. How many wars have you had here in just the past century?" Na yelled defiantly.

"That is irrelevant!" The President shouted.

"I think not." Na replied; "Our records indicate that in the past century earth has had over twenty-five wars here. The last one before the Coonwaadii was in 3567 between Earth and Ppweetan-juthari prime over trade imbalances between your two respective worlds. Can you honestly say that if Mars II were unavailable you would have gone to war twenty-five time? That you would still be able to make war, that your culture, your civilization could withstand that kind of destruction? War is easier now. Though you and your diplomats talk peace, that is all it is, talk. No petition for war space on Mars II has ever been denied, is that right?" Na continued.

"Again I think that is irrelevant!" The President repeated.

"And I think you're both crazy!!" Marsha yelled.

"Agreed!!" All warriors confirmed.

"What?" The President asked.

"Gentlemen, this is no time to debate whether or not this world should exist. The fact remains that it does exist, at least for now. So, while you two continue to fight and bicker about whose right, who's wrong, and debate the philosophical questions of war, alien rights, and all that other bullshit the monitors are on the way here, and in less than two hundred hours, any debate about Mars II will be moot. So, what the hell are we gonna do about them, if anything can be done?" Pageena demanded.

"She's right Mr. Na. Any questions about sovereignty must wait until we have averted the crisis your stupidity has created." The President said gruffly.

"You will get no help from us! Hummf..." Na yelled turning his back to the vid-screen and walking away.

Computer end recording...

In the interest of...

Monitor 614, was making its way back to Mars II along its projected trajectory. At the same time sensors on Kthos, a small planet in the Copernum sector, became active. Scout ships from Kthos, its neighbor Tealifa, and a few colonized moons, were sent to investigate this odd sensor pattern. Though traveling at near hyper-net velocity, the monitor having first been detected at the far reaches of their long range scanners, was reached in only a few moments by the Ktos light crusier Dtombpec.

The Light Star, Bulsusanka, and Happy Farmer joined the Dtompec now converging on the monitor. Light Star was the fastest of the four and could easily outrun most of the others as it took the lead.

"Light Star to Dtombpec. It's only a few meters long and about one and a half in diameter. Looks like a lost probe of something. What should we do?"

"Dtombpec to Light Star. I don't know, its telemetry indicates it's on an assigned course. Doesn't look too dangerous I suggest we leave it alone and let it go. At the rate it's traveling it'll be out of our system in about an hour anyway."

"Dtombpec...we copy. Sounds like a good idea to me. It was a good drill though. Light Star out."

The sixteen crew members of the four ships could not for one instant have understood what would happen next. Monitor 614 began to purge residual anti-matter through its main vents out into a large field of dark matter more than three parsecs wide and eight long. The field density was far greater than elsewhere in the region. Though there was no explosion, the resulting low yield matter anti-matter conversion set up a chain reaction all along the entire dark matter field. Intense verteron and tetrion radiation saturated the area. Anything within the field was immediately converted into a miniature stellar mass.

Now the gigantic cloud moves with the cosmic winds and in its wake Ktos, Tealifa, and all outlying colonies, are left as shining reminders of 614's passing. Eventually, the cosmic winds will disperse this cloud of destruction. That will not however restore the lives of those creatures in their tiny ships, nor the lives of those on their tiny world now burning brightly in the cosmic darkness.

On Mars II Marsha, Pageena and Dafora had been racking their brains for hours trying to come up with a plan to stop the monitors. Campbell, Jarob, and Benni had offered their brain power for a while, such as it was. But they burned out quickly as soon as they began to realize they were in over their head. Na, true to his word, did nothing, and would not allow anyone else, including Nina to assist them. The President and Na continued their argument for another hour or so until both finally realized how deeply entrenched they were in their opposing idioms.

However, most of the kidnapped warriors were allowed if they wanted to, to assist provided they keep the peace, something that proved to be more difficult than one might imagine, under the circumstances.

"Could we construct a Hyper-net?" Marsha asked looking at Dafora.

"Sure, in about eight years." Dafora replied wryly.

"No, I mean a small version. One with just enough field strength to pull in the monitors and direct them to one of the universal deserts, one of these vase black areas here." Marsha replied point to the map.

"But, in order to give it that much field strength all ships within five parsecs would have to be moved or they'd be sucked in too. Not to mention the power consumption something like that would require." Pageena replied.

"Na does this machine have a replicator?" Marsha asked

Na said nothing.

"Look, Na you've got two choices. Either tell her what she want to know or I'll kill you right here and now, with my bare hands if necessary." Campbell bellowed puffing out his massive chest and getting a most menacing look in his eye.

Immediately the guards turned all weapons in Campbell's direction. A move the other warriors took full advantage of and in a matter of seconds, all guards were set upon by the warriors. In the struggle that followed, a random plasma discharge struck Benni in the chest. His eyes glazed as he fell, and with his dying breath he whispered Dafora's name.

"BENNI!!" Dafora yelled.

Pulling her weapon she then grabbed Na, pressing it hard against his temple.

"You bastard! I should kill you right here!" She wept.

"Daf! Daf! Listen to me hon! We all know how you felt about Benni. And we're sorry. But, you have to understand that the function of all warriors is to die. If you kill Na, there's a good chance that we'll all die. Then who'll stop what's happening?" Marsha calmly replied.

"But...but you couldn't know. You couldn't know how I felt." Dafora replied easing her grip on the weapon.

"We know Daf, we know much more than you think we do. Though you did not tell us with words, or perhaps you did not tell him. He knew, as we do. You loved him, didn't you. We could see in your eyes, hear in your voice, and tell by the way you acted when he was near. But, killing Na won't bring him back." Pageena said as she slowly approached Dafora.

"But, you don't understand I...I bear his child." Dafora wept as she again pressed the blaster hard against Na's temple.

As her finger increased its tension on the trigger, Jarob's large strong hand came from behind and pulled upward just as the blaster discharged, taking an inch or two off the top of Na's hair. Dafora collapsed sobbing into Pageena's arms. Jarob then took the blaster and again grabbed Na.

"Warrior Meps!" Campbell yelled "We got no time for you to roll around in self pity. When a Warrior falls, we must respect his memory and wait for another time to mourn the loss. We need all minds clear, and all of the team strong. Now, you either help us or go sulk in the corner. We don't need no babies here."

Dafora looked at Campbell, then at Pageena.

"You carry his child, so he is not dead." Pageena whispered.

"No...no Campbell's right. We have no time for this...Thanks Pageena...But Campbell's right." Dafora replied.

"You are one lucky Coov-ra-danzo." Jarob yelled, increasing his grip on Na; "Now tell your guards to drop their weapons and tell the lady what she wants to know, or I'll burn more than your hair!" He continued, pushing the blaster roughly against Na's back.

"Guards drop your weapons!" Na yelled "But, I don't know nothing about replicators!"

"Forget it I found the replicator anyway." Marsha insisted.

"What do you want the replicator for?" Dafora asked, composing herself and dabbing away her tears.

"Hyper-nets are made using satellites like the monitors. The specifications for those satellites are encoded in our 465 scanners, along with information on every other piece of technology we have. If we link that information to the replicator we could make as many as we need. To set them we just punch in coordinates in the Nalvoron transporter and teleport them to the desired location...Ahhh...you Ok, hon?" Marsha replied.

"No...but I will be." Dafora answered.

"While you're doing that I suggest we begin evac of this planet just to be on the safe side." Jarob commented.

"At last a voice of reason." Pageena added.

"Na, how many are on this planet?" Campbell asked, pressing the blaster hard to Na's ribs.

"Onl...only twenty-five thousand or so." Na replied.

"Twenty-five thousand?" Jarob repeated.

"Campbell the transporter your people developed is it only a sight-to-sight or can you use it as a projected beam?" Marsha asked.

"It can be modified as a projected beam." Campbell answered.

"I think I know where you're headed Marsha. What's its range Campbell?" Pageena asked.

"About 1500 parsecs. But, I don't underst..." Campbell replied.

"Mr. President, are you still on line?" Marsha asked striking the keys of the com panel.

"Yes." He replied.

"Where is the planet you intended to sent the people of Mars II too?" Marsha asked.

"Sector 31, coordinates 015, by 340, by 269. Delta Sigma 29 in Earth quadrant 71. Why?" He replied.

"We've got to evac this planet just to be on the safe side. Since no ships with enough space could possibly get here in time to evac a population of over 25,000. I suggest we use a Nalvoron transporter to beam them straight to the planet. We'll be sending the captured warrior there too. They should be safe until their home worlds can pick them up. It's risky, but we're very quickly running out of options." Marsha suggested.

"We also suggest you move all ships away from Mars II, we're planning on constructing a Hyper-net to detour the monitors into sector Zeta 95." Dafora added.

"Isn't that one of the void areas on the Gailian Map of the

Universe?" He asked.

"Yes... We don't know what will happen to them after that but at least they'll be far away." Marsha added.

A sudden shock wave hitting the ground above them sent people racing in all directions as one photon blast after another came in. Though they were too deep to be directly affected the people in the complex above them were being decimated. Smoke, dust, and debris from the conflagration above began to seep into their underground shelter. More and more blasts began to cut deeper and deeper into the ground above them. Then it came to an abrupt halt, and the vid-screen sprang to life with the image of an aging Coonwaadii staring at them with glowing red eyes glaring.

"I am Amara Sontiina Pocto, Commander of the Coonwaadii Heavy Battle Cruiser GPar, and leader of the Coonwaadii Confederation. We are here to avenge the death of my sons Bosrum and Sokieva. Their life signs terminated as they were aiding you in your investigation. There is no need for feeble excuses, nor will any communication be necessary or accepted. Your task is only to die to avenge their murder." He yelled.

"Amara Pocto, a large battle fleet is emerging from hyper-space in sector 14 and another in sector 6 shall we engage them?" A subordinate interrupted.

"No! Concentrate all fire on immediate target. Destroy anything alive one Mars II first, then we take the galaxy as we should have long ago. If we die in the attempt it will be glorious!" He replied.

"But commander they are more than fifteen kilometers below the surface." His subordinate replied.

"Do not question my orders or you shall be food for the Mzalar when we return in triumph to our home world. Use the boring beams. We'll cut holes down to where our photons will be effective! Now get to it!" He commanded; "Viewer off!"

"Are we just about ready with the teleporter?" Campbell asked.

"Almost there, just a couple more minutes." Marsha replied.

"Well hurry girl! Hurry!" He commanded.

"Hey! If I don't nail down the coordinates just right, a shit load of people are gonna materialize in space somewhere, or maybe a few kilometers above the planet making that first step a long one." Marsha replied.

"I have five monitors closing fast on long range scan." Dafora reported.

"You got an E.T.A?" Pageena asked.

"About fifteen minutes. God, they're moving fast!" She answered.

"Any sign of the Battle fleets, the Waaddi reported?" Jarob asked.

"They're coming from the other way." Dafora replied.

"I've got a coordinates lock! Beginning transporter power up and first stage cycling now." Marsha reported.

"How many can be transported in a group?" Campbell asked.

"I've got a lock on the entire southern quadrant, including, shelters and livestock. Looks like it'll take them all." Marsha reported.

"That's impossible the system isn't designed for a group that large." Campbell replied.

"I had to tie into the alien computer complex just to get the power for the transporter in the first place. I guess, the transporter will go according to its input power." Marsha reported; "Transporting now!"

"Any anomalous readings?" Pageena asked.

"Transport complete...all life functions show green." Dafora replied; "We didn't lose anything."

"Targeting area 2." Marsha replied as again the shock waves from photon charges began pounding their position, stronger and closer than before.

"Target confirmed, set at three kilometers from the last transport point." Marsha replied.

"Beams active." Dafora said as another blast shook the bunker.

"Warrior 6, this is Battle Cruiser Baltimore Road, call sign BR 7, we are reading photon discharges in your area. Are you under attack?"

"No shit Einstein!" Jarob replied

"Our fleet will be in position in approximately four minutes. Who is attacking you?" BR 7 asked.

"The Waadii, somehow they found out that Bosrum and Sokieva were killed, how they found out I don't know. Now daddy's pissed and wants revenge." Jarob replied as another blast tore loose an overhead beam, missing him by inches.

"How did Bosrum and Sokieva get killed. There wasn't even a battle?" BR 7 again asked.

"They had some trouble with Mister Na." Jarob replied.

"Who's Mister Na?" BR 7 asked.

"Look! I've got no time for a round of twenty questions? Yes, it's bigger than a bread box and yes we're in some serious shit here. Are you guys gonna give us some cover fire or what?" Jarob yelled.

"We can't." BR 7 announced.

"What! Why?" Jarob yelled.

"We can't fire without orders from a Commander or higher."

BR 7 replied.

"Ok asshole! This is Commander Marsha Tate. You got that? I'm ordering you to fire on that Waadii ship. But, don't destroy it unless you have to. Suggest you target weapons systems only. We've got an evac underway here, so try to keep them off us for ten or fifteen minutes..." Marsha yelled.

"Are we cleared to transport the remaining colonists?" Dafora asked.

"The only ones left are the ones above us and us down here." Pageena replied.

"Well there's no way we can transport all of us, we'll do the ones topside first." Campbell said.

"Duuuh!...Of course! If we went first we couldn't transport them." Dafora replied wryly.

"Ok...Ok just do it!" Campbell commanded.

"Transport lock initiated. Humm...there's quite a bit of damage and a lot of fatalities, should I try to take only the one's that show life signs?" Dafora asked.

"No take them all. We'll sort it out later." Pageena replied.

"While you're doing that I'll start replicating the satellites we need so we can place them before we go." Marsha replied.

"God! I forgot all about that." Jarob said.

"Feeding data to replicator now." Marsha said.

"Transporter sequence engaged. They're all away." Dafora reported.

"Good...once you have a confirmation, lock on to the coordinates on my panel and begin replication and transport in rapid order." Marsha ordered.

"I have confirmation, switching to your console, process being initiated now. How many do you think we should place?" Dafora replied.

"About a hundred should be more than enough." Marsha answered.

"A hundred? Don't you think that may be too many?" Jarob asked.

"Yeah, That many could suck the planet right in." Campbell added.

"That's a chance we've got to take. We haven't got time for debate now. I'd rather let them suck in the whole planet, than have one of those monitors go off in our system." Marsha replied.

"How long will the replication take?" Pageena asked.

"About six minutes." Dafora reported.

"Ok...get set everyone we should be out of here in about ten minutes." Marsha announced.

"I'm reading a large cluster of monitors heading in." Dafora announced.

"How many?" Jarob asked.

"I can't tell. Looks like hundreds." She replied; "The one's we spotted before have taken station keeping positions in the southern polar region."

"Any chance the Waadii will see them and try to take them out?" Campbell asked.

"There's a chance. But, the monitors are so heavily shielded I don't think the Waadii could do much damage to them." Marsha replied.

"Replication and placement complete. Setting coordinates to

our location." Dafora said.

A low hissing static sound above them grew louder. As the bore beam finally broke through to their position.

"BR 7! I commanded you to fire on the Waadii ship." Marsha yelled striking the com panel.

"Which one. Their projecting false holo images, hundreds of them. Sensors read them as solid, when we do a target lock, they disappear and reappear somewhere else." BR 7 replied.

"Just...just do the best you can." Marsha ordered.

Two high pitched whines accompanied the explosions at the far end of the complex sending them all to the floor, and under the consoles for cover. In quick succession, three more explosions followed each one closing in on their location.

"Is the equipment damaged?" Marsha yelled as she stood up and took a quick account of who remained. No one had been injured, but all were covered with a fine layer of dust.

"We're still on line. I have a coordinates lock on our location." Dafora reported.

"Let's get the hell outta here!" Campbell yelled as more high pitched whines accompanied by the flash and concussion of impacts shattered the surrounding walls.

A large explosion, then another and another again sent them all ducking for cover. Pageena was the first to appear, then Campbell, Jarob, and Marsha.

"Are you all ok?" Pageena asked.

"I think so." Jarob answered.

"Where's Daf?" Marsha asked.

"I don't know she was over by the main power console when

the last volley came in." Jarob answered

Turning they could all see that the main power console was now buried beneath roof supports, bracing, and rubble from the surrounding walls.

"Daf!" Pageena yelled.

Campbell rushed over and began pulling large plastiform conduits, and stones away from the console. Dafora was found under the main console in a safe location. Though she had been knocked unconscious, the only real injury appeared to be a large fragment of splintered metal sticking abruptly out of her left thigh. As they pulled her clear of the debris, she let out a cry of pain.

"Ok! She's clear, where are the others?" Jarob yelled.

"They're dead!" Campbell replied, looking at the bloodied, twisted bodies, of Na, Nina and the others, half buried under tons of rubble.

"Activate that thing and get us out of here! " Jarob yelled.

"Powers gone!" Marsha yelled.

"Is there any way to patch into a source?" Campbell asked

"Got any idea what source we could tie into?" Marsha replied.

"No." Pageena replied looking at the others and seeing the same fear in their eyes, she knew must be in her own.

Beep beep beep went Marsha's bio monitor as it again made its hourly sweep.

"What the fuck is that?" Pageena asked.

"It my bio monitor, damn thing does that every hour on the hour to make sure it's on the right set of tits." Marsha replied.

"Wait a minute."Campbell replied tearing open Marsha battle

top.

"Campbell, this is hardly the time." Pageena replied.

"No, this bio monitor has its own power supply doesn't it?" He asked.

"Yes, but it wouldn't power this console long enough to get us to the planet." Marsha replied.

"Would it get us back to the pod?" Campbell asked.

"It might...it just might...I'll need a minute or two to set coordinates and check power." Marsha replied rushing with the bio monitor to the power control console.

"Don't worry about a power check. The only thing we've got going for us now is, the electromagnetic flux of the dead in here won't allow our friends up there to read anything for now as soon as it all calms down though, they will. When they do a scan if we're outta here, to them we'll be dead. So just attach it, adjust it, and let's go!" Pageena replied.

"Ok it's all set, the satellites will activate in sixteen hours, an auto emergency beacon is in place, and coordinates are set for the pod. Activating...NOW!" Marsha said.

As Marsha activated the beam they were transported to the small rocky outcrop on which the pod cluster rested. Jarob and Campbell assisted Dafora in getting inside.

"Should we contact the Battle Cruisers and let them know we're on our way to rendezvous with them?" Pageena asked.

"No...radio silence until we get off this planet!" Marsha insisted; "The only call we should make is a laser beacon to the drop ship. Once that's done I hope we can shield ourselves long enough for it to come get us."

"Shields on maximum. I hope they don't read our power source beacon active and receiving an acknowledgment from the drop ship.

It'll be here in 9 hours." Pageena replied.

"Nine hours...we could be tiny little pieces by then." Jarob replied.

"Maybe not...If the Coonwaadii commander thinks we're dead there's a good chance they won't bother to look anywhere else. According to their scanners, our last location was in the control center fifteen k under that pile of rubble that use to be a town. If they scan it and don't read any life signs there's a good chance they'll think they got the job done. I hope." Jarob said.

"Just to be safe I'm gonna throw out our EM scattering field so that even if they do pick us up. They won't be able to pinpoint our location. Hope the power holds out." Marsha said.

Hours passed. In that time they managed to eat a little and catch up on some sleep. Since there had been no attack on them, it had been assumed that perhaps Jarob was right and the Cooonwaadii had scanned the debris of the town they assaulted, found no life forms and left. Still, they could see the flash of photon charges in the distance, so they knew the Waddii hadn't given up yet.

Pageena and Campbell worked for some time on Dafora's injuries, finally they managed to get the bleeding under control and stabilize her. To be on the safe side though they placed her in one of the five cryo-tubes, used to keep wounded warriors in stasis, until they could reach proper medical care.

The drop ship arrived and soon began to lift the pod cluster off the surface. Needless to say they were all more than happy to be leaving Mars II. In the time they had been waiting, Marsha could only guesstimate where the monitors were. Using a sensor sweep would have been useless with the EM field in place. Not to mention it would have, in all probability, alerted the Coonwaadii to their position.

Once in orbit, it became apparent the Coonwaadii had not given up. In fact they were continuing to fire on the same location. A few seconds after they had escaped the planet's atmosphere, the last of the monitors arrived.

"This is Warrior 6 calling BR 7. We are clearing the planet. I suggest we all get the hell out of here as fast a possible. The monitors will activate in less than an hour and hopefully before that the Hyper-net will come on line. You copy that?" Pageena said using a sight-to-sight beacon to create a secured communications line with BR 7.

"We copy. But, our order won't allow us to leave with the Coonwaadii still firing on Mars II. Since it's you they're after I suggest we go with all possible speed, report the situation to the nearest station and get some back-up out here. We've got some on the way...but I'm not sure it'll be enough." came the reply.

"It looks like some of it is here now." Marsha replied as she watched her vid-screen and noticed several heavy cruisers entering orbit.

"Yep...looks like." He replied; "Look you're about three hours from the nearest hyper-net. You'd better get going we'll keep things in line here."

"Just remember the hyper-net we've constructed is about a hundred times more powerful than any of the others. If you're still here, you'll get dragged into it when it comes on line. Bear in mind also that those monitors in orbit now, are due to descend in about forty-five minutes. Your window looks like no more than twenty minutes. Good luck Warrior 6 out." Marsha replied.

Marsha and Pageena then set the coordinates for the nearest hyper-net in quad-9.

"Engaging systems now." Marsha reported, as the mains came on. There was no feeling of movement, just the sight of Mars II getting smaller on the view-screen.

Computer recording off...

No time...

The Coonwaadii, continued to pour an additional three to four hundred photon torpedoes into the remains of the city. And into the now empty shelter beneath it.

"Coonwaadii Commander. This is Warrior ship Delphi. We urge you to discontinue your siege. Our scanners have swept the planet and found no life forms in the vicinity you have been attacking. Please cease your attack." The commander said.

"We do not recognize your authority here. Go away or you'll be next." Commander Pocto raged.

"Very well sir. But, it is my obligation to inform you that Mars II is now surrounded by more than one thousand monitors that have been enhanced, and are set to decimate Mars II in less than thirty minutes. In addition, a Hyper-net has been constructed 80,000 kilometers off your port bow. Its purpose is to pull all monitors and possibly even Mars II into a region of space now void on our charts, hopefully when the anti-matter charges in the monitors go off, our area will not be affected. These are the facts. The Monitors are due to impact in 29 minutes, 14.35 seconds. We suggest you make for the nearest Hyper-net. This is your final warning!" The Delphi Commander warned.

"I SAID GO AWAY!!!" Commander Pocto replied, as he continued to lob photons at Mars II.

"Very well, But you have had your warning. Delphi out!"

The commander then ordered all ships to vacate the area at once.

"Well be at hyper-net in two hours, forty-two minutes." Pageena reported as she turned and looked at the others. "Jarob you look like shit." She continued.

"This ain't been no walk in the park babe." He replied with a stretch and a yawn.

"How long before the monitors go?" Campbell asked.

"About twenty minutes or so." Marsha replied.

"And when will the hyper-net go on line?" Jarob asked.

"If it works, about five minutes before the monitors are due to descend." Marsha answered.

"Cuttin' it kind of close ain't it?" Campbell responded rhetorically.

"Just working with what we had." Pageena replied; "First rule of Warrior training right?"

They sat in silence and watched the vid-screen, zooming in occasionally to see the progress of the heavy cruisers as they began their evac and closer to see what the Waadii were doing. Not ones to give up easily, the Waadii were continuing the pounding of Mars II. Every minute or so, they would receive cross talk from the cruisers and coded messages from the Waadii ship.

"Should be any second now, the hyper-net is set to go on line in 3...2...1...? What happened? It should have activated?" Marsha replied excitedly.

"Four minutes, ten seconds and those monitors go in." Pageena reported.

"I don't understand it?" Marsha replied.

"Maybe the Waadii attack knocked out the power link to the net." Jarob suggested.

"They're all self generating. Otherwise they wouldn't be able to maintain station." Marsha answered.

"Maybe the timing got thrown off." Pageena replied.

"That's more likely it. Each one is off by a second or so. The phased timing sequencer hasn't got them all up to the same time yet.

Damn!" Marsha replied.

"Can we get any more speed outta this thing?" Jarob asked.

"She's running max now. Hyper 9 is about as hot as she gets." Campbell replied.

"It won't make any difference. If the anti-matter in those monitors goes off in this sector. Everything for 40,000 parsec's or more, will be gone anyway." Pageena replied.

"Anything we can do?" Campbell asked.

"Not unless you're a big fan of praying." Marsha responded.

"A minute, six and it'll all be over." Pageena reported.

The Coonwaadii were reloading when the hyper-net came on line. Seconds later the ship, crew, and Commander Pocto, began to feel the tremendous draw as it began to pull them in. Mars II began to slip and shift as it too, became effected by it. The commander ordered all mains on line and full power. They had waited too long, the taste of revenge so sweet upon his lips, began to sour as faster and faster his ship closed in on the hyper-net.

Any anguish, either commander Pocto or his crew might have felt would be a question of debate for another time. For now though the race between the monitors, and the hyper-net was on. Though some monitors were caught up in the pull of the hyper-net and taken inside. Others began to make their way toward the target.

Marsha felt helpless as she saw the monitors begin their decent on Mars II. All any of the others could do was sit and watch the vidscreen intensely as Mars II moved toward the net. For a few brief seconds, the impacts of the monitors appeared to have little effect, if any.

"Maybe Na fucked up this too?" Pageena replied.

Then a small anomaly in the northern region of Mars II, began to grow. Like a distortion in the air or a strong wind it expanding outward, further and further until the entire planet was involved. As it grew, all color appeared to fade. Then darkness like a hole in space a gigantic maw growing larger, sucking up everything. Tree, mountains and everything part of Mars II was eaten by this voracious beast as it continued to grow to twice Mars II's size, then three times, then four. There was no light, no sound, no sign of disturbance. Only the darkness that rolled toward them.

The vid-screens could only see the ships now and the massive black curtain they were trying to outrun. As each in turn was overtaken, em pulses knocked out all electronic gear. Seconds later as if crushed in the teeth of some mighty demon, the deformation of the ship began until it resembled an old tin can. As they watched the ships being pulverized into smaller and smaller bits, soon they were too small to see. But still the blackness rolled on.

"Oh..God!! Looks like we're gonna catch a piece of it." Marsha screamed; "Hold on! I'm shutting down all systems!"

Pageena grabbed Campbell, Jarob grabbed Marsha. Each couple then shared a chair, and strapped in. They held each other tightly, eyes closed, as the blackness surrounded them. Fortune smiles on the unworthy sometimes as the darkness did little beyond taking out the lights and caused shorts sending sparks and smoke in all direction, on all decks. A second bang and jolt had the same effect. Followed by a third, fourth and fifth, with each round causing the occupants to grip each other with greater fervor.

As the last of the jolts subsided, each of them opened their eyes and looked at each other. With that look that says "What happened??"

"This is impossible!" Marsha said as she unhooked the strap securing her and Campbell in place; "I'm gonna do a systems check."

"I'll help you." Pageena replied following Marsha's lead.

A system check revealed there to be little damage to life support, replicators, vid-screens and archives, and the computer. There were however, sporatic communications, maybe two or three seconds in hyperdrive, a small impulse deck with maneuvering thrusters. The hyper-net they were making is over 90 parsec's away and could take years to reach if at all. Using the hyperdrive they managed to get about half way home. But there were still months of travel to go before they could reach even the outer systems.

"So, now we drift, use the impulse deck sparingly to conserve fuel even though it is re-generative. It doesn't make any difference. Maybe in a couple of years from now we'll be able to reach hyper-net 12. A new galaxy has formed in the void where once only darkness reigned. The chances are good that it will be an anti-matter system. Even at hyper 25 it would take more than six thousand years to get to it however. Still, it will make some scientists go into an orgasmic frenzy, as they watch it develop, grow and reveal to someone how our little bang, compares with the big one, that in theory created the universe in which we now find ourselves adrift.

In the meantime, each of us writes our own version and chapter to this chronicle. Using the archives, transmissions, personal logs, and other sources, as well as any recollections we may have. We thought it would be proper to record these events so we can fill the days and months ahead. We know we're not stranded forever in this darkness and we know that sooner or later we'll be found.

The cryo-tubes come in handy, four can sleep a month or two while one remains to look after things. Daf is in one all the time, both because of her injuries, and due to her pregnancy. If we can delay gestation a while, maybe we'll get help before she delivers. Our hopes don't rely too heavily on that however.

We write, sleep, eat and sometimes when hope is at its lowest, turn to each other for comfort. But, it's my turn now, to write, to relive, and relate. Each word that is written is also transmitted via an old style carrier wave radio system we managed to knock together. Kind of like a note in bottle.

Jarob should be waking soon. I don't know if he will have more to add to this or not. Seems ironic to me that he went from one prison to another. Maybe someday we'll talk about it. Maybe not, Marsha Tate going off line now.

Warrior 6 Complement:

Mars II

Jarob Ka   Pageena Timbo   Lucifer Campbell
Dafora Meps   Benni Jack Taylor   Marsha Tate

Year of Red Moon Rising - 3997

END UPLINK....END TRANSMISSION

Made in the USA
Middletown, DE
13 March 2015